Why Do Buildings Collapse in Earthquakes?

Why Do Buildings Collapse in Earthquakes?

Building for Safety in Seismic Areas

Robin Spence
Emeritus Professor of Architectural Engineering and Fellow of Magdalene College, University of Cambridge, United Kingdom

Director, Cambridge Architectural Research Ltd, United Kingdom

Emily So
Reader in Architectural Engineering and Fellow of Magdalene College, University of Cambridge, United Kingdom

Director, Cambridge Architectural Research Ltd, United Kingdom

This edition first published 2021
© 2021 John Wiley & Sons Ltd

Registered Offices
John Wiley & Sons, Inc., 111 River Street, Hoboken, NJ 07030, USA
John Wiley & Sons Ltd, The Atrium, Southern Gate, Chichester, West Sussex, PO19 8SQ, UK

Editorial Office
9600 Garsington Road, Oxford, OX4 2DQ, UK

For details of our global editorial offices, customer services, and more information about Wiley products visit us at www.wiley.com.

Wiley also publishes its books in a variety of electronic formats and by print-on-demand. Some content that appears in standard print versions of this book may not be available in other formats.

Library of Congress Cataloging-in-Publication data

Names: Spence, R. J. S. (Robin J. S.), author. | So, Emily, (K. M.), author.
Title: Why do buildings collapse in earthquakes? : building for safety in
    seismic areas / Robin Spence and Emily So.
Description: Hoboken, NJ : Wiley-Blackwell, 2021.
Identifiers: LCCN 2020053851 (print) | LCCN 2020053852 (ebook) | ISBN
    9781119619420 (hardback) | ISBN 9781119619451 (adobe pdf) | ISBN
    9781119619468 (epub)
Subjects: LCSH: Buildings–Earthquake effects. | Earthquake resistant design.
Classification: LCC TH1095 .S65 2021 (print) | LCC TH1095 (ebook) | DDC 693.8/52–dc23
LC record available at https://lccn.loc.gov/2020053851
LC ebook record available at https://lccn.loc.gov/2020053852

Cover Design: Wiley
Cover Image: © Professor Mauricio Beltran

Set in 9.5/12.5pt STIXTwoText by Straive, Pondicherry, India

C9781119619420_080721

Contents

Acknowledgements

We have been generously supported in the preparation of this book by many friends and colleagues, who have read and commented on parts of the text, provided suggestions and illustrations and have been a valuable source of ideas and inspiration over many years.

We would like first to thank Yutaka Ohta whose dedication to earthquake risk reduction and foresight has been a constant motivation to us both. We would also like to acknowledge the late Laurie Baker and the late Paul Oliver, whose understanding of vernacular architecture and its value has been an inspiration for us.

In the preparation of this book, we greatly appreciate the comments and advice on particular chapters from Andrew Coburn, Antonios Pomonis, Svetlana Brzev, Edmund Booth, Mary Comerio and Rashmin Gunasekera.

We appreciate also the cooperation of Rajendra and Rupal Desai, Amod Dixit, Lucy Jones Toshitaka Katada, Randolph Langenbach, Tracy Monk and Edward Ng in reviewing our profiles of them and their work.

We thank our colleagues Hannah Baker and Weifeng Victoria Lee for help in the preparation of maps and diagrams used, as well as for reviewing parts of the manuscript, Yue Zhu for her help with preparation of a number of the images used, and Sandra Martinez-Cuevas for supplying the diagram of fragility curves used in Chapter 5, based on her analysis of the Cambridge Earthquake Impact Database (CEQID).

We especially appreciate the painstaking work of Charlotte Airey in developing the building construction type and damage mode drawings used in Chapters 5 and 6.

We also greatly appreciate the contributions made by the global panel of experts who responded to the surveys of national successes and failures reported in Chapter 8. An acknowledgement to them is given at the end of that chapter. We would like to acknowledge the help of our colleagues at EEFIT, to Allan Brereton and the committee for their help in making available appropriate illustrations. And to our colleagues at Cambridge Architectural Research and the Department of Architecture at the University of Cambridge for their support of the book.

Finally, we would like to express our personal thanks to our families. *From Emily*: my gratitude to Alex and Clara for being the perfect lockdown husband and baby during this challenging year. *From Robin*: my thanks as always to Bridget for her unfailing support and wise counsel through this, and so many projects in the past.

1

Introduction: Why This Book

1.1   Earthquakes – An Underrated Hazard

Earthquakes have been a threat to human habitation throughout history, but until relatively recently, their causes were poorly understood. In the pre-scientific era, they were commonly ascribed to divine intervention. By the time of the Lisbon earthquake in 1755, there were many who understood that earthquakes had natural causes, but the mechanism remained unexplained, and the supernatural explanation was widely proclaimed, especially from church pulpits (Udias and Lopez Arroyo 2009). And over 150 years later, according to observer Axel Munthe (1929), the inhabitants of Messina, destroyed by a massive M7 earthquake in 1908, cried 'Castigo di Dio' ('punishment from God').

Only with the development of plate tectonics in the twentieth century has it become understood that earthquakes are associated with active faults in the earth's crust, with most of the largest occurring at the boundaries of the tectonic plates as they interact with each other (as explained in Chapter 4). We can now identify with some precision whereabouts on the earth's surface large earthquakes will occur. From measurements of the movements at plate boundaries, and from the historical record, we can make estimates of the largest magnitude event which can occur on a fault section, and approximately, the frequency with which events of different magnitude will occur. But the largest events commonly have return periods of several centuries or more (Bilham 2009), and science is still unable to predict, even to within a few decades, when the next large earthquake on any fault section will occur.

There is some evidence that the global earthquake mortality rate (deaths per 100 000 of the world's population) has been rather gradually reducing over the last century or so. But it is a very slow rate of improvement, and the variation from decade to decade is very large. The first decade of the twenty-first century was a bad one, with several earthquakes resulting in more than 50 000 deaths. Yet, over the same timescale, death rates from many other causes, such as infectious diseases and road accidents, have been very significantly reduced (ourworldindata.org/causes-of-death 2020). This has been made possible with the introduction of public health programmes and protection measures, backed by government legislation and action programmes, but supported and implemented by the general public. Such programmes could similarly be applied to reduce earthquake risk, but in many countries most at risk, this has not so far happened. Why is this?

Why Do Buildings Collapse in Earthquakes?: Building for Safety in Seismic Areas,
First Edition. Robin Spence and Emily So.
© 2021 John Wiley & Sons Ltd. Published 2021 by John Wiley & Sons Ltd.

The greatest impact from earthquakes is nearly always the damage to buildings (and other built artefacts – roads, buildings, dams) from the ground shaking caused by the propagation of the earthquakes' waves through the earth's crust, which can result in destruction over a wide area. Over the twentieth century, understanding the nature of ground motion and the way in which this is transmitted through structures has enabled engineers to develop ways to build buildings which are able to withstand the expected ground shaking with limited damage. This understanding, gradually increasing through the development of structural engineering theory and practice, combined with detailed field investigation of the effects of successive earthquakes has enabled codes of practice for building design to be developed, and these are nowadays mandatory for new construction in most cities of the world.

But, as the world's population grows, and urbanisation increases in pace, there are many places where new buildings are being constructed without any reference to good engineering practice for earthquake resistance.

This is partly because those responsible for constructing the new buildings are unaware or possibly unconcerned that a large earthquake may occur any time soon, and building controls are lax. It is also due to lack of education, information, skill and sense of urgency on the part of builders and building owners (Bilham 2009; Moullier and Krimgold 2015).

In rural areas of many poor countries, buildings are largely constructed using highly vulnerable materials such as adobe and unreinforced masonry. Poverty and lack of understanding, combined with a vast demand for new dwelling places, are thus fuelling the creation of a series of future disaster scenarios (Musson 2012).

In order to understand why buildings collapse in earthquakes and to find out what we can do about it, we must look at each of the three ingredients of the problem: earthquakes, buildings and people.

1.2   Earthquakes, Buildings, People

One of the reasons why earthquake risk does not get acted on is because it is not well understood by the public. Although the likely locations of large earthquakes are now known, the timescale of their recurrence is very long, and for most people at risk the last occurrence of 'the big one' for which they need to be prepared is many centuries ago, often before the present cities existed. People may be aware that they are living in an earthquake zone but fail to appreciate the possibility of events much larger than recent experience. In 2008, a modelling exercise, the California Shakeout, was done to support earthquake protection action for Southern California, which is threatened by a large earthquake on the San Andreas Fault (Jones and Benthian 2011). Lucy Jones, who led the modelling team speaks of the 'normalisation bias, the human inability to see beyond ourselves, so that what we experience now or in our recent memory becomes our definition of what is possible'. Seismologists had identified much greater earthquakes in the past than those in recent memory, but the last great earthquake on that section of the San Andreas Fault was in 1688. The modelling exercise, based on a plausible, but by no means worst-case scenario magnitude 7.8 earthquake on the southern section of the San Andreas Fault, showed that around 1500 buildings would collapse, and 300 000 would be severely damaged, causing around

1800 deaths and $213 billion losses. Fires would break out and could become uncontrollable. And the disruption caused to roads and pipelines would cause massive disruption to business, lasting for months. This modelling exercise led to a huge public awareness and preparation programme which has resulted in much reduced risks in California over the past decade.

But considerably more devastating consequences face many of the growing cities in other earthquake zones, particularly in Asia. The southern edge of the Eurasian Plate, stretching from the Mediterranean to China, and including Myanmar and Indonesia, is responsible for 85% of the world's historic earthquake deaths. And this is a region in which cities are today growing rapidly both in size and in number, fuelled by global population rise and urbanisation. Seismologist Roger Musson points to the risk in Tehran, today a city of 12 million people. The last major earthquake on the North Tehran Fault, passing close to the city centre, was in 1834 at a time when Tehran was a small town: an earthquake of $M > 7$ hitting Tehran today could cause as many as 1.4 million deaths. And seismologist Roger Bilham (2009) has estimated that a direct hit on a megacity (>10 million population) somewhere in the world once a century is now statistically probable, with a possible death toll exceeding one million, because of the combination of hazardous locations and structural vulnerability. The World Bank estimates that three billion people will live in substandard housing by 2030. By 2050, the UN projects that two-thirds of the world's population, around 7 billion people, will live in urban areas.

Unfortunately, because the threat to each city is seen as remote, protection from earthquakes is given a lower priority than other issues. Few households prioritise spending on safety from future earthquakes above pressing immediate concerns, like providing extra space or better comfort, unless required to do so by regulation. And elected governments tend to look for expenditure programmes and new regulations which will give returns within their current tenure of office, despite evidence that money spent on disaster mitigation often avoids much greater losses over time. For this reason, general development expenditure is given priority over disaster risk mitigation. And even within that part of government budgets devoted to natural disasters, those from other natural hazards are often given priority. Windstorm and flood damage are more immediate risks, particularly as these are becoming worse as a result of climate change.

Optimistically and opportunistically, the climate change agenda has provided a global focus on resilience of communities to natural threats. It is recognised that especially in developing countries, cycles of disasters have depleted decades of progress made in development. The deaths and destruction from earthquakes are preventable. Whilst the hazard itself is natural, the disasters are largely man-made, and completely preventable with proactive interventions.

1.3   The Authors' Experience of Earthquake Risk Assessment

The overall aim of our work over four decades at the University of Cambridge's Department of Architecture and at Cambridge Architectural Research Ltd has been to understand the vulnerability of buildings to earthquakes globally, in order to estimate the damage which is likely to occur from future earthquakes. This knowledge can be used to provide a sound basis to improve the building stock, and reduce damage, loss of life and disruption from future earthquakes. We have developed our knowledge of building vulnerability through a series of collaborative research projects, supported by the European Union and the UK Government and Research Councils, and through work for individual cities, companies managing portfolios of buildings and insurance companies. But the primary source of our knowledge and experience of buildings' behaviour in earthquakes has been post-earthquake field missions. We have been involved in EEFIT, the UK's Earthquake Engineering Field Investigation team, since it was founded in 1982, and have between us participated in field missions in Japan, Italy, Turkey, India, Pakistan, Peru, Indonesia, China, New Zealand and the South Pacific. The detailed nature and aims of these field missions are discussed in Chapter 2: but an essential element in all cases is to describe and document the types of building affected and the types of damage observed.

Successive projects have examined in detail the problems of particular regions. In the 1980s, we examined the traditional stone-masonry construction of rural Eastern Turkey and conducted shake-table tests in Ankara to investigate simple ways to reduce their vulnerability, the cause of many deaths in earthquakes of the previous decade. In the 1990s, we investigated the options for protecting historic European cities such as Lisbon and Naples from likely future earthquake damage, and we looked at the performance of buildings which had been strengthened following previous earthquake damage. We also developed a method for assessing human casualties from earthquakes based on the level of building damage, and with colleagues in New Zealand applied this to the city of Wellington.

Since 2000 we have worked with others to develop loss modelling approaches to estimating damage and casualties, on a city-scale (in EU collaborative projects), for insurance companies, or with the US Geological Survey, for rapid post-disaster damage assessment. And we have applied our knowledge to assist organisations with large portfolios of buildings to identify those which should be upgraded.

We have also worked with teams developing new ways to assess earthquake damage using remote sensing, and led the team developing the Earthquake Consequences Database (So et al. 2012) for the Global Earthquake Model (GEM). And we have applied similar approaches to assessing vulnerability and damage to buildings from other natural hazards such as windstorms and volcanic eruptions. All this work is described in detail in technical project reports and published papers, referred to in the chapters which follow.

1.4 Aims of This Book

The title of this book asks a question: Why do buildings collapse in earthquakes? In exploring the many layers of the answer to this question, and the many answers in differing contexts across the world, we want to demonstrate that this is not just, not even primarily, a technical question, but also a social, organisational and even political question. In this book, we look at buildings not only as assemblages of materials and components put together to achieve certain functional ends, but also as products of a society and a culture. We aim to explain the physical reasons why buildings fail to withstand earthquakes, but also to attempt to understand the social, economic and political reasons why earthquake disasters continue to happen. And through this combined understanding, we want to point to the actions that can be taken to improve seismic safety, and identify who should be taking them.

With this aim, we hope to reach a wider audience than those interested in the purely technical aspects of earthquake protection, who would prefer a non-mathematical approach to the subject, with limited technical detail. Thus, the book is designed to be read by all those interested in the consequences of earthquakes, or concerned for their own safety as occupants of buildings in earthquake areas. It is also intended for those who have responsibility for ensuring the safety of others in earthquakes, whether as government officials, political representatives, building owners or managers of businesses. The book is written for a non-technical readership, but will also be of interest to all those professionally involved in disaster preparedness and earthquake engineering, as well as to students and practitioners of architecture and engineering seeking a broad overview of the consequences of earthquakes for buildings.

Some readers of the book will live in an earthquake zone, in which case they will want to know if their homes or workplaces are vulnerable, and what they can do to protect themselves from an earthquake, in advance or when it happens. Other readers may own or manage buildings in earthquake zones, or be responsible for the safety of those who occupy them; they will want to know what steps they as owners might be able to take to provide adequate safety. Other readers may be responsible, as architects and engineers, for the design of new buildings or the refurbishment of older ones in earthquake zones and will want to know what the essential steps in building for safety in such areas are. Yet, others may have a more general interest in natural disasters and need an informed but largely non-technical account of how buildings have performed and of how the way today's buildings are constructed has been influenced by past earthquakes. The book aims to provide useful and accessible answers for all of these groups of readers.

1.5 Outline of the Book

The remainder of the book is divided into eight chapters. Chapter 2 presents field evidence of how buildings behave in earthquakes. It discusses how post-earthquake field investigations have contributed to our understanding of building behaviour. It gives brief accounts of 10 of the most significant earthquakes of the past 20 years. It concludes with an assessment of the overall trends of earthquake damage and casualties over time, and their distribution between richer and poorer countries.

Chapter 3 looks at how buildings are constructed in the world's most earthquake-prone regions. It considers first how the local climate affects local patterns and traditions of building, and shows how those traditional building forms affect earthquake performance. The world's areas of the greatest earthquake risk are then subdivided into 10 separate zones, and the patterns of building typical of each are described and illustrated, distinguishing rural and urban types.

Chapter 4 explains what causes earthquakes, and shows how the ground motions caused by them are felt by buildings and how buildings respond. It also considers other ways in which earthquakes can affect buildings through ground deformation, landslides, tsunamis and fire outbreaks, and points to the growing risk of compound disasters triggered by earthquakes.

Chapter 5 considers how buildings of different types of construction respond to the principal earthquake hazard of ground shaking. It classifies buildings into their different types and subtypes according to the main material of the load-resisting system – masonry, reinforced concrete, timber and steel. For each main type, it describes the typical behaviour in an earthquake from the onset of damage to collapse, based on field observations. And it suggests cost-effective ways in which each type of building could be made more earthquake-resistant. Chapter 5 also compares the earthquake vulnerability of different building types, showing the wide disparities that exist within the global building stock.

Chapter 6 looks at human casualties caused by earthquakes. It identifies the main causes of casualties, and how these relate both to building performance and to occupant behaviours. It shows how the expected number of casualties from a particular earthquake can be estimated for loss modelling, using either statistical or engineering approaches.

Chapter 7 considers different routes by which the earthquake resistance of buildings can be improved. It looks first at the engineering design of buildings and how codes of practice are used to achieve acceptable safety levels, both in the construction of new buildings and in the strengthening of existing buildings, and discusses associated costs. It also considers limitations in the effectiveness of building control regulations and implementation of codes of practice, and describes how building for safety programmes have been used to improve the construction of non-engineered building in poorer countries.

Chapter 8 reports on a global survey of the successes and failures of earthquake protection, country by country, based on responses from 39 experts in 28 different countries. For each responding country, the identified successes and failures are examined, and the countries are divided into three groups 'high achievers', 'limited achievers' and those with 'continuing and growing risks', indicating the wide disparity of performance across the world.

What is technically possible will only be achieved by the action of individuals and society as a whole, and its institutions. Thus, Chapter 9 concludes the book with an examination of what part different organisations and groups of people can play in meeting the overall challenge of earthquake protection. The separate roles of governments (national and local), non-government organisations (NGOs), the scientific and professional community, businesses, homeowners and individual citizens and the insurance industry are considered, and suggestions are made for ways in which each group could act more effectively.

Emphasising the message that it is ultimately the action of individuals that counts, the book contains a series of profiles (located as boxes within the appropriate chapters) of some individuals – 'game-changers' – whose actions have made a notable contribution to earthquake protection in their particular situation. These advocates show what we can do with the knowledge to build safe buildings before an earthquake strikes and to stop preventable deaths. Earthquakes are an underrated hazard: but by ensuring safe buildings and earthquake awareness before the earthquakes strike, we can make the threat unremarkable.

References

Bilham, R. (2009). The seismic future of cities. *Bulletin of Earthquake Engineering* 7: 839–887.

Jones, L. and Benthian, M. (2011). Preparing for a "big one": the great Southern California shakeout. *Earthquake Spectra* 27: 575–595.

Moullier, T. and Krimgold, F. (2015). *Building Regulation for Resilience: Managing Risks for Safer Cities*. Washington, DC: GFDRR, The World Bank.

Munthe, A. (1929). *The Story of San Michele*. London: John Murray.

Musson, R. (2012). *The Million Death Quake*. Palgrave Macmillan.

So, E.K.M., Pomonis, A., Below, R. et al. (2012). *An Introduction to the Global Earthquake Consequences Database (GEMECD)*. Lisbon: 15 WCEE.

Udias, A. and Lopez Arroyo, A. (2009). The Lisbon earthquake of 1755 in Spanish contemporary authors. In: *The 1755 Lisbon Earthquake Revisited* (eds. L. Mendez-Victor and C. Oliveira). The Netherlands: Springer.

2

How Do Buildings Behave in Earthquakes?

2.1 Learning from Earthquakes

When a large earthquake occurs, it causes human casualties, damages buildings and infrastructure, and affects livelihoods, society and the wider economy. It also sets in motion a process of relief and recovery, damage assessment and then rebuilding, carried out by governments, NGOs, commercial firms and individual households. It is important that the experience of each earthquake is recorded in detail, and that the lessons learnt are identified and passed on, both for the benefit of the affected country in its attempt to improve preparation for subsequent earthquakes, and also for the international community. Much of the damage caused by an earthquake is visible only for a short time, because demolition and rebuilding often start within a few days, so it is important that damage investigations start rapidly after an event. But it is equally important that, if they are to be useful for international comparison, such investigations should be done in a systematic way.

The need for speedy but systematic post-earthquake investigations has led to the formation of a number of international earthquake reconnaissance teams whose aim is to be available for rapid deployment after an earthquake. They are composed of earthquake specialists from different disciplines, and generally include team members from the affected country. Each team conducts a survey whose exact scope depends on the scale and type of damage. But the study generally includes investigations of the seismological and geotechnical aspects of the event, the damage to buildings and to infrastructure, and the way in which relief and rescue has been conducted. On return, the team produces a report which is available to all who are interested, and is commonly made available on openly accessible websites. The team also communicates the findings through various technical meetings.

The Learning from Earthquakes programme of the California-based Earthquake Engineering Research Institute (EERI) has the most experience of such field reconnaissance missions, and has conducted more than 150 investigations since it began after the 1971 San Fernando, California earthquake. In the United Kingdom, the Earthquake Engineering Field Investigation Team (EEFIT), working in conjunction with the UK's Institution of Structural Engineers, has conducted more than 30 investigations since its

Why Do Buildings Collapse in Earthquakes?: Building for Safety in Seismic Areas,
First Edition. Robin Spence and Emily So.
© 2021 John Wiley & Sons Ltd. Published 2021 by John Wiley & Sons Ltd.

formation in 1982 following the Irpinia (Italy) earthquake of 1980. Similar organisations exist in several other countries (Spence 2014). The cumulative findings of the missions have been very influential in formulating research programmes which have studied aspects of the physical damage, response and recovery from multiple events. And these research programmes in turn have led to steady improvements of national and international codes of practice for building, as well as assisting in understanding the vulnerability of different types of affected facilities and in developing ways to enhance earthquake safety internationally (EERI 1986; Spence 2014).

Both authors have been involved with several EEFIT post-earthquake reconnaissance missions. Our direct knowledge of the types of buildings affected in earthquakes, and our understanding of their behaviour, is largely derived from these earthquake missions, as well as from some more detailed field investigations and household surveys carried out independently. The following sections give brief accounts of 10 of the most significant earthquakes of the last 20 years, partly based on our own observations, but also making use of the field reports of our colleagues in the EERI and EEFIT teams and other reports. As we are concerned in this book primarily with buildings, these brief accounts emphasise in particular the range of building types which were affected and the levels and types of damage caused, topics which we will return to look at in more detail later in the book. They also touch, where appropriate, on the methods of damage investigation used.

Table 2.1 lists the most significant events of the twenty-first century up to 2018. It includes all those events which, according to the EM-DAT database, killed more than 4000 people, and also all those which had a damage cost exceeding US$3.9bn. The 10 events briefly described here include the 9 events with the highest casualty tolls of the last 20 years, and one other event, the New Zealand Christchurch event of 2011. This was particularly significant not for its casualties, which were relatively low, but for the very high financial cost of the damage caused, and for its particular impact on the historic masonry buildings of the city of Christchurch.

The chapter concludes with some general observations about earthquake damage, an assessment of global damage trends and the distribution of damage between different regions and country groups. In this way, we aim to approach an assessment of the question: how well are we, as an international community, doing in trying to limit the effects of earthquakes for this and future generations?

2.2 Significant Earthquakes Since 2000

2.2.1 The 26.1.2001 Bhuj Earthquake: Mw7.7, 13 481 Deaths

At 8.46 a.m. on 26 January 2001, India's 52nd Republic Day, one of the most devastating earthquakes ever to strike India occurred in the Kutch Region of Gujarat State. The earthquake of moment magnitude Mw7.7 and focal depth 23 km was located approximately 70 km east of the historic city of Bhuj. Heavy ground shaking affected an area of tens of thousands of square kilometres, although there was no surface fault rupture observed. The isoseismal map prepared by the EERI team indicates that the area subject to shaking at a level exceeding MM Intensity VIII ('heavily damaging') was over 30 000 km^2 (Jain et al. 2002).

Table 2.1 Significant earthquakes worldwide since 2000, ordered by number of deaths.

Date	Country	World Bank Income Group	Event	Magnitude (Mw)	Total deaths	Total damage (US$bn)	Insured losses (US$bn)	Percent insured
26/12/2004	Indonesia, Thailand, Sri Lanka, India	UM, LM	Indian Ocean earthquake and tsunami	9.1	225841	7.8	0.48	6.2
12/01/2010	Haiti	L	Haiti	7.0	222570	8	0.2	2.5
12/05/2008	China	UM	Wenchuan	7.9	87476	85	0.37	0.4
08/10/2005	Pakistan	LM	Kashmir	7.6	73338	5.2	0	0
26/12/2003	Iran	UM	Bam	6.6	26796	0.5	0	0
26/01/2001	India	LM	Bhuj	7.7	13481	2.6	0.1	3.8
11/03/2011	Japan	H	Great Tohoku[a]	9.1	>18000	210	37.5	18
25/04/2015	Nepal	L	Gorkha	7.8	8831	7.1	0.1	1.4
26/05/2006	Indonesia	UM	Yogyakarta	6.3	5778	3.1	0.04	1.3
28/09/2018	Indonesia	UM	Sulawesi[a]	7.5	4340	1.5	0	0
21/05/2003	Algeria	UM	Boumerdes	6.8	2266	5	0	0
03/08/2014	China	UM	Yunnan	6.2	731	5	0	0
27/02/2010	Chile	H	Maule[a]	8.8	562	22	8	36
19/09/2017	Mexico	UM	Puebla	7.1	369	2.9	1.3	45
24/08/2016	Italy	H	Amatrice	6.2	296	7.9	0.12	2
20/04/2013	China	UM	Lushan	6.6	198	6.8	0.023	0
22/02/2011	New Zealand	H	Christchurch	6.1	181	15	12	80
16/04/2016	Japan	H	Kumamoto	7.0	49	20	5	25

(*Continued*)

Table 2.1 (Continued)

Date	Country	World Bank Income Group	Event	Magnitude (Mw)	Total deaths	Total damage (US$bn)	Insured losses (US$bn)	Percent insured
23/10/2004	Japan	H	Niigata	6.6	40	28	0.76	3
16/07/2007	Japan	H	Niigata	6.6	9	12.5	0.34	3
20/05/2012	Italy	H	Emilia-Romagna	6.0	7	15.8	1.3	8
14/11/2016	New Zealand	H	Kaikoura[a]	7.8	2	3.9	2.1	54
04/09/2010	New Zealand	H	Darfield	7.0	0	6.5	5	77%

Income groups are from World Bank data (High, H; Upper-middle, UM; Lower-middle, LM; Low, L). See also Table 2.2. Dates are given in DDMMYYYY format. Some casualty and loss data are amended based on more recent estimates (Pomonis 2020).

[a]Events with significant tsunami impacts are shown.

Sources: CRED (2020); Pomonis, A., 2020. Personal communication.

The area has experienced a previous large earthquake (Mw about 8.0) in 1819, and a moderate Mw6.1 one in 1956, and is in the zone with the highest earthquake loading requirements in the Indian code of practice for the design of buildings.

Load-bearing masonry is the predominant way of building throughout the affected area, but methods have changed over time. The most common masonry technique is a single-storey house with walls of random rubble stone masonry set in a mud mortar, with a clay tile roof: these buildings are found everywhere, both in the main towns and in the villages (Figure 2.1). More substantial dwellings use dressed or semi-dressed stone or sometimes clay brick walls; these are commonly two-storey buildings. In recent years, the use of rein-forced concrete (RC) slabs for floors and roofs, with coursed masonry walls, has become common in the wealthier parts of Kutch (Figure 2.2). The main towns have also significant numbers of multistorey apartment blocks in RC (Figure 2.3). None of these forms of build-ing were spared by the intense and widespread ground shaking.

The major city of Gandhidham, and four large towns Bhuj, Anjar, Bhachau and Rapar, all in the Kutch district, were devastated, as was every village within a wide area. Over 230 000 one- and two-storey masonry buildings and several hundred concrete frame build-ings collapsed. However, as pointed out by Sudhir Jain (2016), the collapse rate of buildings in the zone of highest intensity was much lower in this earthquake than in the 1993 Mw6.2 Latur earthquake in India's Maharashtra province where rubble stone walls with heavy mud roof are typical.

In Ahmedabad, about 200 km from the epicentre, severe shaking was experienced and over 100 multistorey RC frame buildings collapsed. A survey of damaged buildings in Bhuj and neighbouring villages by EEFIT (2005), including the author's team, showed that the rubble masonry buildings performed worst (over 30% collapse rate) while masonry with RC slabs and RC frame apartment buildings performed better (7 and 3% collapse rates). The collapse of buildings in Ahmedabad, all of which were of multistorey RC frames, can be

Figure 2.1 Stone masonry building in the Kutch district damaged in the Bhuj earthquake.

Figure 2.2 Brick masonry building with reinforced concrete floors damaged in the Bhuj earthquake.

Figure 2.3 Reinforced concrete building in the Kutch district and typical damage patterns. Notice ground floor failure.

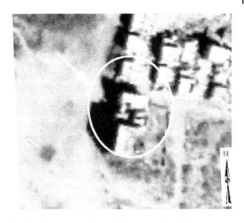

Figure 2.4 Damage to a reinforced concrete building in Bhuj, and view of the same building in satellite image, arrow showing viewing direction. *Source:* Saito et al. (2004).

attributed to amplification of the ground motion through the deep alluvial deposits on which Ahmedabad stands, coupled with poor design and construction – soft-storey apartment blocks were common. The Indian earthquake design code in use at the time is well-written and comprehensive, but it was not binding on private builders, and was largely ignored (see Chapter 8).

This earthquake was one of the first in which available high-resolution satellite imagery could be used to identify the damage to individual buildings, and this resulted in several studies to develop this technology (Saito et al. 2004) (Figure 2.4).

The final toll of dead and injured shows that altogether 13 481 people were killed in the earthquake (Jain et al. 2002). There were more than 166 000 injured, 20 000 of them seriously. From medical reports, it is clear that both death and injury were mainly the result of traumas associated with the collapse of buildings. Over 1000 school students and teachers were killed, though because it was a public holiday many schools were closed. There were also more adult female than male deaths (Murty 2005). There can be little doubt, though, that failure of weak masonry walls and the resulting collapse of dwellings was the main cause of death, and the magnitude of the death toll is a reflection of the very wide area over which heavy ground shaking was observed, combined with the weakness of the typical masonry buildings.

2.2.2 The 26.12.2003 Bam Earthquake: Mw6.6, about 27 000 Deaths

This earthquake occurred at 5.26 a.m. local time, on a hitherto unidentified fault passing under the historic city of Bam (Berberian 2005). Surface ruptures were identified along this fault south of Bam, and extending northwards towards the centre of the city. The area as a whole is one with a well-known history of active seismicity (nine earthquakes have been felt in Bam since the beginning of twentieth century). The earthquake was catastrophic in the city of Bam itself (far more than would be expected for an earthquake of this magnitude), as well as in the nearby town of Bharavat and neighbouring villages of Kerman Province. The intensity map, produced by IEES, indicated heavily damaging shaking intensity over an area of about $1000\,km^2$. The degree of damage in the city was clearly visible in high-resolution satellite images (Figure 2.5).

The massive death toll in Bam has been attributed to the extreme weakness of the adobe houses which are inhabited by the majority of the population. This method of building has been documented in the World Housing Encyclopedia (Maheri et al. 2005).

Adobe construction is an appropriate response to the climate of Southern Iran, given high day–night temperature swings, and also the lack of timber available for construction. At the time of the earthquake, this was still the predominant way of building in Bam. But in the event of an earthquake its weakness is extreme. The problems include (Maheri et al. 2005):

- Thick heavy walls, which attract large lateral seismic forces.
- Lack of connections between perpendicular walls.
- Heavy domed or vaulted mud roofs, exerting lateral pressure on walls.
- Poor quality of the adobe units (local sun-dried mud).
- Poor quality of mortar and bonding.
- Lack of foundations.
- Limited maintenance.

Architectural conservator Randolph Langenbach who made his own study in Bam following the earthquake (Langenbach 2015) has suggested that the use of straw reinforcement in adobe construction may have allowed termite attack, which could have reduced the inherent cohesion of the material. Certainly, many of Bam's adobe buildings simply disintegrated as a result of the ground shaking, leaving only heaps of dried mud brick rubble (Figure 2.6). The danger to occupants was increased by their close spacing, leaving little opportunity for escape, and this also inhibited search and rescue. Since the earthquake attempts have been made to develop a way of building dwellings which conforms to the climatic and space requirements, and uses local materials, but which is able to resist earthquakes (Maheri et al. 2005).

The huge death toll of nearly 27000 was about 25% of the population of Bam at that time. It was undoubtedly the result of the collapse of very large numbers of adobe dwellings,

Figure 2.5 High-resolution satellite imagery of the centre of Bam taken (left) before and (right) after earthquake, clearly indicating the extent of the damage. *Source:* Satellite image ©2021 Maxar Technologies.

Figure 2.6 Failure of adobe dwelling in the Bam earthquake. *Source:* World Housing Encyclopedia. Reproduced with permission of EERI.

coupled with the early morning time of day, when most people were sleeping. It has been reported that only 2% of those who died were in buildings which did not collapse (Ghafory-Ashtiany and Mousavi 2005). Of the 23 600 injuries, 9477 were serious, and had to be treated in hospitals in Kerman and elsewhere as all the hospitals in Bam were severely damaged. Building collapse-related traumas constituted most emergency surgery cases. However, it has been suggested that a further very significant contribution to the death toll was the lack of immediate response capability (Movahedi 2005). The local emergency response capability was totally destroyed by the earthquake, and for the crucial first 24 hours the only rescue was being carried out by the local survivors using their bare hands. The loss of electricity meant that rescue stopped at nightfall, and freezing temperatures reduced the chances of overnight survival under the rubble. Asphyxiation resulting from the huge amount of dust was suggested as a further cause of many deaths (Movahedi 2005).

2.2.3 The 26.12.2004 Indian Ocean Earthquake and Tsunami: Mw = 9.1, 225 841 Deaths

At 7.59 a.m. on 26 December 2004, one of the largest earthquakes of the last 100 years anywhere in the world occurred in the Sunda trench in the Indian Ocean. At Mw9.1, the earthquake was one of the largest ever recorded, and had the longest duration of faulting ever recorded (between 8 and 10 minutes), with a fault rupture extending for 1300 km. The earthquake caused ground shaking over a wide region, but because of the extraordinary length of the fault rupture and the movement on it, the earthquake also triggered a massive and destructive tsunami, which devastated the coasts bordering the Indian Ocean, causing huge loss of life. The initial ground shaking was destructive throughout Aceh Province of Indonesia, particularly in the main city of Banda Aceh, and also in the Andaman and Nicobar Islands. But the tsunami carried the earthquake's energy over a much wider region, causing destruction throughout coastal northern Sumatra, and in all the countries bordering the Indian Ocean. Casualties caused by the tsunami were reported in 12 different countries, but most of the tsunami-related deaths occurred in Indonesia (165 000), Sri Lanka (36 000), India (16 000) and Thailand (8000). In Aceh Province of Indonesia, it destroyed virtually every village, town, road and bridge along a 170 km stretch of coast that was not more than 10 m above sea level. The death toll was over 16% of the entire population of the northernmost six districts of the province. Inundation depths reached up to 20 m in parts of Sumatra, 5–8 m in Thailand, and 2–5 m in South-eastern India and Eastern and Southern Sri Lanka (EEFIT 2006).

The tsunami was devastating to small buildings wherever the inundation depth was 2 m or more, and huge numbers of buildings of timber or traditional masonry were destroyed in Indonesia, Thailand and Sri Lanka (EEFIT 2006) (Figures 2.7 and 2.8). RC buildings of several storeys often survived but with serious damage, although there were cases of collapse through scour under the foundations. The huge loss of life was primarily due to the direct effects of the tsunami itself. Victims were either drowned directly or as a result of injuries caused by impact with debris from buildings or other objects: 'falling structures and waters full of swirling debris inflicted crush injuries, fractures and a variety of open and closed wounds' (WHO 2006). Tens of thousands were swept out to sea, and were ultimately recorded as missing, and were presumed drowned.

Figure 2.7 Damage caused by the 26 December 2004 tsunami at Unawatuna, Sri Lanka where the inundation depth was about 5 m. Damage to a masonry building. *Source:* EEFIT. Reproduced with permission.

Figure 2.8 Damage caused by the 26 December 2004 tsunami at Unawatuna, Sri Lanka where the inundation depth was about 5 m. Damage to a reinforced concrete frame building. *Source:* EEFIT. Reproduced with permission.

In Sri Lanka and Thailand, many of the victims were foreign tourists. It has been estimated that the death rate in the worst hit areas in Sri Lanka and Thailand was over 10% of the resident population within 1 km of the coast. A study of the life loss in Indonesia, India and Sri Lanka found that in certain locations the disaster claimed four times as many lives among women as men (EEFIT 2006).

It is clear from all accounts that an effective warning system, coupled with a better understanding of the phenomenon of tsunamis among both residents and visitors could have saved many lives, since the tsunami struck the Thai and Sri Lankan coasts more than 90 minutes after the earthquake.

A study carried out by our own team of the experiences of eyewitnesses (Spence et al. 2009) showed very strong correlation between survival and distance from the shore: all of those within 15 m of the shore reported serious injury or fatalities in their group, but less than half of those more than 30 m away did. Most survivors who were in the affected zone attributed their survival either to prompt action in moving to safer ground, or to being in a building which survived.

2.2.4 The 8.10.2005 Kashmir Earthquake in Pakistan: Mw = 7.6, 73 338 Deaths

On 8 October 2005 at 8.50 a.m. local time, an earthquake of magnitude Mw7.6 struck the Kashmir regions of Pakistan and India. The epicentre was a little north of Muzaffarabad, the major town of Pakistan's AJK (Azad Jammu and Kashmir) Province. It was located on the Jhelum Thrust (Tapponier et al. 2006), part of the well-established thrust fault system associated with the subduction of the Indian plate below the Eurasian plate.

Heavy ground shaking was felt over a very wide area, and was devastating for the nearby towns of Muzaffarabad, Balakot, Bagh and Rawalakot; but damage was severe in towns up to 50 km away, including Murree, Abbotabad and Mansehra in Pakistan, and Uri and Baramulla in India. In the three worst-affected districts of AJK (Muzaffarabad, Bagh and Rawalakot), 84% of buildings were damaged or destroyed. In Islamabad, 100 km away, a recently built multistorey RC apartment block collapsed (EERI 2005, 2006a) killing 74. As many as 5000 school buildings were seriously damaged or destroyed killing 18 095 students and 853 teachers (Asia Development Bank 2005; EEFIT 2008a). This was the most destructive event in the Indian subcontinent in the last 50 years, causing as many as 75 000 deaths and 68 700 serious injuries in Pakistan, and 946 deaths and 4400 injuries in India (EEFIT 2008a). Altogether, about 450 000 homes were destroyed, and over 2.8 million people were left homeless.

The high death toll was undoubtedly primarily due to the widespread collapse of buildings in the area, most of them of masonry. Because of the harsh climate, buildings have traditionally been made from thick stone masonry, often using rounded riverbed stones in poor-quality mud mortar, with steel sheet or thick mud roofs (Figure 2.9). In the past, such walls were often tied together with timber lacings and the roof independently supported on timber columns. However, timber is less and less used because of its scarcity and high cost, and the severe ground shaking would have been more than enough to cause overturning or disintegration of the walls, followed by roof collapse. In many places, more modern building

Figure 2.9 2005 Kashmir earthquake: damage in the Muzaffarabad district. Aerial view of damage to traditional residential buildings. *Source:* EEFIT. Reproduced with permission.

Figure 2.10 2005 Kashmir earthquake: damage in the Muzaffarabad district. Damage to reinforced concrete construction.

types using concrete blocks and RC frames also collapsed (Figure 2.10), and this included many government-built schools and barracks. Evidence from post-earthquake field investigations showed poor-quality building standards (EEFIT 2008a).

A factor which certainly also contributed to the high death toll was the inaccessibility of much of the mountainous affected area, as a result of the numerous landslides triggered by the earthquake blocking roads. The emergency services were thus very slow to arrive, and many of the survivors had to walk long distances in difficult terrain to reach a functioning health centre; this also complicated injuries, bringing on infections and resulting in more drastic medical measures. Many more victims with head and chest injuries from falling masonry did not survive until medical help arrived. Unfortunately, search and rescue capability in the crucial early stages was overly concentrated in Islamabad, where few buildings failed, rather than being sent to the epicentral area (So 2009).

There were no official data from which causes of death could be established, but a survey of survivors in 500 families in one of the worst affected neighbourhoods in Muzaffarabad was carried out by our team in conjunction with University of Peshawar (So 2009) to establish patterns and causes of death and injury. This confirmed that, whatever form of construction was used, the major cause of serious injury and death was structural collapse resulting in entrapment.

2.2.5 The 27.5.2006 Yogyakarta Earthquake: Mw6.3, 5778 Deaths

Centred on the densely inhabited Yogyakarta region in eastern Java, this earthquake occurred at 5.53 a.m. local time, with its epicentre in Bantul district. The area of high ground shaking intensity was greater than $200\,km^2$. More than 156 000 houses and other structures were destroyed, killing 5778 people and seriously injuring more than 40 000. The loss of housing accounted for more than 50% of the total damage, and it has been suggested that the death toll could have been much larger had the earthquake not occurred at a time when many were awake and involved in household tasks outside their houses. Nearly all the deaths and serious injuries which did occur were as a result of the collapse of buildings (So 2009).

The typical house in the affected rural areas (katcha house) is a single-storey masonry building, with burnt clay brick or concrete block laid in a weak cement or lime mortar (Figure 2.11). The roof is of timber or bamboo trusses and rafters, supporting timber battens and clay tiles. There are no connections between the roof members and the walls. A minority of more recently built houses (since 1990) are built using confined masonry walls, with RC columns and beams, as described in Chapter 5, but with traditional roof structures. A few are of timber frame construction (EERI 2006b).

The lack of adequate ties between roof and walls, and the lack of out-of-plane strength of the walls were responsible for the collapse of many of the katcha houses. The buildings built using confined masonry generally performed better, though the earthquake exposed many failures in jointing and reinforcing such structures (EERI 2006b).

As in the 2005 Kashmir earthquake, there were no hospital data available to establish causes of death and injury, but a survey of survivors in 523 families in 4 of the worst affected districts, sampling 2652 individuals, was carried out by our team in conjunction with Yogyakarta's Gaja Mada University (So 2009) to establish patterns and causes of death and injury. This confirmed, as in the companion survey in Kashmir, that the major cause of

Figure 2.11 Collapse of house of traditional construction, Yogyakarta earthquake. *Source:* Boen (2016). Reproduced with permission.

serious injury and death was structural collapse resulting in entrapment, with no clear difference between different types of construction. However, in this event, a high proportion of the building occupants moved outside at the onset of the earthquake (67% of those in buildings which collapsed), and this may have saved many lives.

2.2.6 The 12.5.2008 Wenchuan Earthquake: Mw7.9, 87 476 Deaths

The Mw7.9 Wenchuan earthquake occurred at 14.28 local time on 12 May 2008, with its epicentre in Wenchuan county in Sichuan Province of China, at a depth of 19 km. It caused a fault rupture of around 240 km along faults which form the boundary between the Longman Shen mountains to the north-east and the Sichuan basin to the south-west. This is a densely populated region, and the earthquake was devastating to a large area, affecting more than 250 000 km^2 and 30 million people (EERI 2008). Peak ground acceleration exceeding 0.5 g was felt over a wide area. The earthquake resulted in approximately 87 500 deaths (including 17 920 missing people), and 375 000 injuries, and required almost 1.5 million people to be relocated. It was the most lethal earthquake to strike China since the Mw7.5 Tangshan earthquake in 1976 which killed an estimated 242 000 people.

Although most of the deaths were caused by the collapse of buildings, a notable feature of this earthquake was the very large number of slope failures, causing landslides, rockfalls and mudflows. By one estimate there were more than 15 000 such failures, which resulted in around 20 000 deaths, or nearly 30% of the total (Yin et al. 2009).

Most of the buildings affected by the earthquake were masonry buildings. In the rural and mountainous areas, these were traditional unreinforced masonry (URM) buildings of brick or stone with timber floor and roof structures, which were highly vulnerable to earthquake ground shaking. In the urban areas also many URM buildings existed, often of several storeys, with precast concrete hollow-core floors and roofs, and this included many school buildings (Figure 2.12). In the urban areas, one common form of building had a RC ground floor with URM above, while others were of confined masonry construction (EERI 2008). The urban URM buildings did not perform well in the earthquake and there were many partial or catastrophic collapses.

Many of the mixed RC and brick constructions too were seriously damaged, including many ground floor 'soft-storey' collapses. There were some recently constructed RC frame buildings in the larger towns and these were reported to have performed relatively well, and to have resisted collapse in spite of the fact that the ground shaking was much more severe than the levels required by the design codes over most of the area (EEFIT 2008b).

Particularly noteworthy was the tragic collapse of many school buildings. It has been estimated that more than 7000 school classrooms collapsed in the earthquake, and, as children were in school at the time of the earthquake (14.28 local time), that possibly more than 10 000 children died as a result. Often, collapsed school structures were sited next to other relatively undamaged buildings. This led to angry protests by bereaved families, whose only child was often the victim. Protesters blamed both shoddy construction and the government which should have supervised such construction more effectively. Official investigations took place and are said to have identified many design defects, and there have been subsequent changes to the required standards for school construction (Wong 2008).

Figure 2.12 Partially collapsed five-storey masonry middle school building in YingXiu in the Wenchuan earthquake. *Source:* EEFIT. Reproduced with permission.

2.2.7 The 12.1.2010 Haiti Earthquake: Mw7.0, Estimated More Than 222 000 Deaths

The Mw7.0 earthquake which struck the Republic of Haiti at 16.53 local time on 12 January 2010 was one of the most destructive in history. It had its epicentre 25 km from the capital city of Port-au-Prince, at a depth of 13 km, and very strong to severe ground shaking was felt throughout the city, which had a population of 3 million, and the surrounding region, resulting in collapse or critical damage to more than 300 000 homes. In addition, the Government of Haiti estimated that 60% of the nation's administrative and economic infrastructure was lost, and 80% of the schools and 50% of the hospitals were destroyed or damaged. The death toll was initially given by the Haiti Government as 316 000 (DesRoches et al. 2011) but an estimate of around 220 000 is now widely accepted, although other estimates range from 46 000 to 159 000. This would still mean that death toll as a proportion of the nation's population was greater than in any earthquake in modern times (DesRoches et al. 2011).

Although detailed post-mortem data are not available, all reports point to poor standards of building construction leading to building collapse as being the principal cause of the immense death toll (DesRoches et al. 2011; EEFIT 2010; Marshall et al. 2011). The majority of houses were of one or two storeys, made of a mixture of concrete block masonry and RC frame construction. In some cases, the frame was built after the masonry walls, as in confined masonry construction; in other cases, the walls were infill within a previously built frame, and some investigators found that the 'wall-first' buildings performed better in the

Figure 2.13 Typical damage to low-rise informal building in Port au-Prince. *Source:* EEFIT. Reproduced with permission.

earthquake. But for both types of construction, the materials used, the amount and detailing of the reinforcement and the process of construction were generally inadequate, resulting in buildings which were totally unable to resist the earthquake forces they were subjected to. Failures of the concrete frame leading to the overturning of masonry walls were common (Figure 2.13). Engineered commercial buildings mostly performed better, and most of the few traditional timber-framed buildings remaining in the city suffered only moderate damage (EEFIT 2010).

The vulnerability of Haiti's building stock had a number of underlying reasons. There had been no significant earthquakes in the area since the eighteenth century, so earthquake awareness was low. Haiti is the poorest country in the Western hemisphere, and lacks effective government institutions, so there was no effective building code. Topography and ground conditions probably contributed to the scale of the damage, with some soils causing ground motion amplification, and failure of buildings on steep slopes resulting from foundation failure (EEFIT 2010).

Because of the extent of damage and the chaotic post-disaster situation, investigating and quantifying the extent of the damage presented a serious problem in this event, and it was the first in which a crowdsourcing approach was used to map the damage using high-resolution satellite images (Ghosh et al. 2011). Over 600 individual damage analysts from 24 different countries participated in this work. A first map of building damage, identifying over 5000 collapsed buildings, was assembled within 48 hours of the project being commissioned, and a second map, using higher-resolution aerial photographs, and delineating more than 30 000 heavily damaged or destroyed buildings, was completed within three weeks (Ghosh et al. 2011). These maps made an important contribution to the World Bank and EU's Post Disaster Needs Assessment, completed in the first few weeks after the event.

Ground surveys were conducted by an EEFIT team to validate these maps (EEFIT 2010), and found that while the damage they identified was correct, the aerial maps missed a significant amount of serious damage, because it could not be seen in the vertical image.

2.2.8 The 22.2.2011 Christchurch New Zealand Earthquake: Mw6.1, 181 Deaths

The earthquake occurred at 12.51 p.m. local time, causing severe shaking throughout the City of Christchurch, the largest city in New Zealand's South Island. The Mw6.1 earthquake was an aftershock of the Mw7.1 Darfield earthquake which took place five months earlier, and had also caused shaking and damage in Christchurch, but the epicentre of the 22 February 2011 event was much closer to the city, and caused ground shaking well in excess of the design level which the New Zealand Code specifies for Christchurch (EEFIT 2011a). The ground shaking destroyed hundreds of buildings in the city, including many old URM structures as well as two large office buildings, and 181 people were killed, nearly three quarters of them in the two collapsed office buildings. The earthquake also caused extensive liquefaction of soft alluvial ground both in the Central Business District and the eastern suburbs which added to the extent of building damage. The estimated total losses were US$15bn, of which around 80% was insured (King et al. 2014).

For more than a century, most residential buildings in New Zealand have been timber-framed buildings with lightweight roofs and cladding, and most of these buildings survived the earthquake with little damage except to masonry chimneys and wall claddings (EEFIT 2011a). However, in liquefaction-affected areas, the foundations of many residential buildings failed, resulting in their subsequent demolition. Many RC buildings in the Central Business District were seriously damaged, which was to be expected, given the high level of ground shaking. This shaking damage was combined with foundation damage caused by liquefaction, and many of the damaged buildings were subsequently demolished. The two major office buildings which collapsed were both built to pre-1976 design requirements. The most widespread damage was to URM buildings, which comprised the historic core of Christchurch, including its cathedral, built in the 1870s to a design by Sir George Gilbert Scott, where the spire collapsed. All the masonry buildings in Christchurch had been built before the 1930s, and the materials used, the structural arrangements and the wall-to-wall and floor connections were inadequate for this level of shaking (EEFIT 2011a). There was widespread failure, mostly through overturning of poorly connected walls and gables (Figure 2.14), some of which killed and injured pedestrians.

New Zealand has had programmes of strengthening URM buildings since 1968 (see Chapter 8), and there were a significant number of URM buildings in Christchurch which had been retrofitted. In general, these performed better than the non-retrofitted buildings, but many of these too were damaged, and have had to be demolished. By 2013, more than 90% of the non-retrofitted URM buildings, and over 70% of the retrofitted URM buildings in the Central Business District (CBD) had been demolished (Moon et al. 2014). The earthquake was therefore an important test for the efficacy of previously used retrofitting techniques.

Detailed damage assessment of all the structures in the affected area was carried out by engineering teams in the weeks following the earthquake, for assessing occupant safety (Galloway et al. 2014). Thus, aerial imagery, although available, was not needed for damage

Figure 2.14   Typical damage to pre-1930s masonry buildings in Christchurch.

assessment as it was in Haiti. The crowdsourcing approach to damage assessment using remote sensing pioneered in the Haiti earthquake (Ghosh et al. 2011) was, however, able to be tested in this event by the authors' team against extensive ground-truthing (Foulser-Piggott et al. 2015). This study confirmed that the aerial image-based damage identification rates are quite low. For example, for masonry buildings in the CBD, only 56% of those buildings given a red ('unsafe') or yellow ('restricted-use') tag by the ground inspection teams were identified as damaged from the remote-sensing imagery.

Many important lessons have been learned from this well-studied event, about building performance, retrofitting, insurance, damage assessment and recovery procedures, as well as the limitations of life-safety-based codes of practice. These lessons are now beginning to be implemented, not just in New Zealand, but worldwide (Chapter 8).

2.2.9   The 11.3.2011 Great Tohoku Japan Earthquake: Mw9.1, Over 18000 Deaths and Missing

Less than three weeks after the Christchurch earthquake, a huge Mw 9.1 earthquake occurred at 14.46 local time in the Japan trench off the coast of Tohoku in north-east Japan. The event had the largest magnitude recorded in Japan since the beginning of instrumental seismology around 1900. Intense and damaging ground motion, lasting as much as four minutes, was felt over most of northern Japan, and the earthquake triggered a tsunami, which was damaging over about 650 km of the coastline of four regions (Sanriku, Miyagi, Joban and Kanto). The tsunami toppled sea defences, inundated more than 500 km^2 of land and destroyed many settlements and towns along the coastline. The effects of the tsunami also led to the failure and release of radioactive material from two reactors of the Fukushima

Daiichi nuclear power station, requiring a massive and prolonged evacuation of hundreds of thousands of residents from within 20 km of the site.

Direct losses from the event are thought to be in excess of $215bn, making this the world's costliest disaster, and indirect losses from the nuclear reactor failure continue to rise. The total casualties resulting from the event have been estimated over 18000. The vast majority (>90%) of these were deaths by drowning, directly caused by the tsunami, but significant numbers of casualties (>4%) were crush injuries caused by the collapse of buildings over a wider area. As in Christchurch, the scale of the damage was largely because the buildings and the sea defences were not built to withstand the effects of an earthquake of that magnitude. At that time, the maximum magnitude envisaged in the Japan trench, on which building and coastal defence design was based, was Mw8.4.

The damage caused by the earthquake and tsunami was studied by several reconnaissance teams. The summary of damage to buildings presented here is largely based on reporting by the UK's EEFIT team (EEFIT 2011b). The building stock in the 20 towns and cities affected by ground shaking (but not the tsunami) which were visited by the EEFIT team consisted of an estimated 81% timber-framed structures, 14% RC structures and 4% steel framed structures. In surveys conducted by EEFIT in the areas where the ground shaking was seriously damaging, as many as 25% of all buildings

Figure 2.15 Typical damage in tsunami-affected region. Timber frame building in Kamaishi City at a location with a maximum of 7 m inundation. *Source:* EEFIT. Reproduced with permission.

Figure 2.16 Typical damage in tsunami-affected region. Steel frame building in Yamamoto-cho, Miyagi Province which was struck by a tree trunk in the debris flow. *Source:* EEFIT. Reproduced with permission.

were recorded as unsafe and another 30% as damaged (EEFIT 2011b). Each of the three major building types was equally affected, though damage to RC buildings was concentrated on those built before improved earthquake-resistant design regulations were introduced in 1981.

In the coastal areas affected by tsunami inundation, damage was immense, but extremely variable, depending on inundation depth, flow velocity, debris entrainment in the flow and sheltering. Inundation depths varied up to a maximum of 16 m at Onagawa town, where most buildings were destroyed, and 12% of the population living in the inundated zone were killed. The structural frames of RC and steel frame buildings generally survived, except where impacted by heavy debris, but damage to cladding and non-structural elements was extensive (Figures 2.15 and 2.16). Timber-framed buildings were, however, frequently destroyed and sometimes swept from their foundations by the force of the tsunami at inundation depths of 3–4 m or more.

Overall, data collected by the National Police Agency (NPA) in November 2011 (NPA 2011) showed that 120 157 buildings were totally destroyed and a further 830 000 buildings had been damaged.

2.2.10   The 25.4.2015 Gorkha Nepal Earthquake: Mw7.8, 8831 Deaths

This Mw7.8 event occurred on 25 April 2015 at 11.56 a.m. local time, with an epicentre near the town of Gorkha, 80 km west of Nepal's capital city Kathmandu, and with a focal depth of 19 km. It was followed by an intense series of aftershocks over several months, with their epicentres spread over a length of 200 km east and west from the main shock. The largest of these aftershocks, on 12 May, had a magnitude of Mw7.3, and three others had magnitudes greater than Mw6.0. The earthquakes resulted from ruptures on the Himalayan arc, a highly active fault zone associated with the subduction of the Indian plate to the south under the Eurasian plate to the north, the process which is responsible for the building of the Himalayan mountain range. The cumulative casualties and damage from this sequence of events was immense. Over half a million houses were destroyed, as were many government and heritage buildings, as well as roads, bridges and water supply systems. Damage was increased by many landslides, and rescue was hampered by the inaccessibility of many affected villages. The Nepal Government's post-disaster needs assessment report (EEFIT 2015; EERI 2016; Government of Nepal 2015) estimated that there were over 8790 deaths and 22 300 serious injuries caused by the earthquake, spread over 14 districts. The death toll subsequently rose to an estimated 8831 in Nepal. The cost of the disaster has been estimated to be around US$7bn.

Observations on building construction and damage are taken from the reports of the EEFIT and EERI reconnaissance missions (EEFIT 2015; EERI 2016). According to these reports, in both urban and rural areas of Nepal, masonry construction predominates. In the urban areas, many buildings are of three to four storeys in height, and built in close proximity to each other. Masonry is constructed using fired brick, which is laid in a lime or mud, sometimes cement, mortar. Floors and roof structures were traditionally of timber (floor planks on joists), and in better-built houses these were connected by pegs to wall plates to prevent overturning (EEFIT 2015). More recent masonry buildings have RC floor and roof slabs. Urban areas also have a proportion of RC frame buildings, with masonry infills. Few are built according to the Nepalese code of practice or guidelines for concrete buildings, and there is no inspection (EEFIT 2015).

In rural areas masonry construction is low-rise, using local stone or brick, with a lime or mud mortar. As in urban areas, floors have timber beams and roofs use timber beams or trusses covered with a lightweight tiles or metallic sheet.

Although better-built structures survived the earthquake with minor damage, the earthquake exposed many weaknesses in building practice. Typical damage to masonry buildings included the separation of outer and inner layers of masonry walls because of poor connection within the walls, out-of-plane overturning of walls because of poor connections at corners and poor wall-to-floor connection (Figures 2.17 and 2.18). Many of the casualties were caused by the overturning and collapse of masonry walls. The performance of RC buildings was better than that of masonry buildings, and the majority of those observed by EEFIT, both in Kathmandu and throughout the affected area, showed little or no damage (EEFIT 2015). However, the EERI team (EERI 2016) noted many deficiencies leading to damage or collapse, which they attributed both to problems with building configuration – large overhangs, trapezoidal plans and soft storeys, as well as to poor construction standards. Pounding of adjacent structures with insufficient gaps was also a common source of damage.

Many important heritage buildings, including temples, were seriously damaged or destroyed by the earthquake. There was evidence though, that retrofitting of some traditional masonry buildings using wall ties had been effective. And the work of the National Society of Earthquake Technology, NSET, in strengthening school buildings, and in creating guidelines for improving earthquake safety of rural buildings, was considered by EEFIT (2015) to have protected many buildings from serious damage (see Chapter 7).

Figure 2.17 Typical damage to masonry construction in urban areas of Nepal. *Source:* EEFIT. Reproduced with permission.

Figure 2.18 Typical damage to masonry construction in urban areas of Nepal. *Source:* EERI. Reproduced with permission.

2.3 What Can We Learn from These Significant Earthquakes?

The 10 earthquakes described in the preceding sections account for 99% of all the recorded earthquake deaths worldwide between 2000 and 2019. In two of these events (the 2011 Tohoku and the 2004 Indian Ocean event), the earthquake-triggered tsunamis were responsible for most of the deaths. In another one (the Wenchuan earthquake), ground failures (landslides and rockfalls) were responsible for perhaps 20% of the deaths. But the evidence of these events confirm earlier findings based on events up to the 1990s (Coburn and Spence 2002) that the collapse of buildings has been by far the largest cause of death.

The descriptions also tell us that in cases where collapse of buildings is the predominant cause of death, most of the buildings which collapsed were masonry buildings. Many different methods of masonry construction were tested in these earthquakes, and not all were found equally vulnerable. Relatively good performance was noted in buildings of well-constructed confined masonry, and also those of brick or block masonry where the walls were tied together with RC floors and roof slabs. Particularly, poor performance was noted in buildings of adobe masonry (2003 Bam earthquake), and buildings using rubble stone masonry, in which the walls were prone to disintegration, bringing down heavy roofs (Nepal, Kashmir and Bhuj). But brick and block masonry buildings also performed poorly (in the Yogyakarta and Wenchuan earthquakes, for example) when walls and roofs were not adequately connected. The Haiti earthquake tested a type of masonry/RC frame mixture, which has some similarity with confined masonry, but did not meet the essential requirements of that form of construction, and many occupants were killed by the collapse of such buildings. In Christchurch too, much of the damage was related to the older masonry buildings in the old city centre, though nearly three quarters of the 181 deaths happened because of the collapse of two mid-rise RC office buildings.

In the Tohoku (Japan) and the Christchurch (New Zealand) earthquakes, the residential buildings were mostly single-family one or two-storeyed timber-frames, and the evidence showed that these buildings survived the severe ground shaking well. There was damage to brick elements, like chimneys and cladding panels, but few buildings collapsed.

In most of these events, a part of the building stock, particularly in the cities, was built using RC frame construction, for multistorey apartment buildings and for commercial or mixed commercial/residential buildings. These were tested in the Bhuj, Kashmir, Nepal and Wenchuan earthquakes, and many such buildings collapsed in each of those events. In all of these countries, there are codes of practice which define how to design RC buildings to withstand seismic action, but it was evident to the reconnaissance teams that such codes were often not being implemented, and that the failings noted in previous earthquakes (inappropriate building form, poor materials, poor detailing of reinforcement) were being repeated. The Bhuj and Kashmir earthquakes also provided further evidence that tall RC buildings can be at risk at a considerable distance from the earthquake's epicentre. Hundreds of tall buildings collapsed in Ahmedabad situated more than 200 km from the epicentral areas of the Bhuj event. The Margalla Towers apartment block collapsed in Islamabad situated 100 km from the epicentral areas of the Kashmir event. RC buildings built before current codes of practice were introduced were also heavily damaged or collapsed in the Tohoku and Christchurch events.

The two events in which a major tsunami occurred (2004 Indian Ocean and 2011 Tohoku) have provided new evidence for the resistance of different forms of construction to the forces of tsunami waves. Most of the deaths in these two events were not caused by the collapse of buildings, but by drowning. Nevertheless, huge numbers of buildings were destroyed by the tsunami, causing immense economic loss. In both the Indian Ocean and the Tohoku tsunami, buildings of traditional construction (timber frame or masonry) were destroyed, and sometimes swept from their foundations when the inundation height at the shore was 2 m or more. RC and steel frame buildings were also seriously damaged by the force of the water and the impact of the debris (often from other destroyed buildings) carried by the incoming or retreating wave, though their structural frames generally survived. Some interesting cases of engineered buildings being overturned have also been thoroughly investigated in order to update tsunami design guidelines (Macabuag et al. 2018).

Some of the events provided tests for retrofitting programmes carried out in previous years. In Nepal, the retrofitting programme for school buildings carried out in the previous 10 years was found to be very effective as explained in Chapter 7. In New Zealand, however, the effectiveness of retrofitting carried out under the earthquake risk buildings programme was mixed, and numerous previously retrofitted buildings were damaged in the earthquake and subsequently demolished (though not necessarily as a direct result of earthquake damage).

In general, the events demonstrated that where buildings are built in accordance with codes of practice adopted internationally since the 1980s, they performed well, even in places (Wenchuan and Christchurch) where the level of ground shaking was extreme and considerably higher than the code allowed for. But they also showed that implementation of the codes is often patchy, either because they are not required by law or because they are poorly implemented or because many buildings are constructed informally. This applied in particular to school buildings which collapsed in large numbers in several earthquakes (Kashmir, Wenchuan, Haiti and Nepal) confirming already growing evidence of the high vulnerability of this vital public service.

In some cases (Bam, Kashmir and Haiti), the number of casualties was reported to have been magnified as a result of very limited or delayed emergency response, because of infrastructure damage, difficult terrain or simply because the local emergency response infrastructure and personnel were also directly affected. And, in both the events in which large tsunamis were triggered, a lack of public awareness and a lack of an adequate warning system was considered to have increased the death toll, although in the case of Japan it is considered that warnings and pre-event evacuation plans which did exist (though based on lower tsunami inundation level) did contribute significantly in the reduction of loss of life.

More detailed accounts of the performance of buildings are given in the EEFIT and EERI reports referenced in relation to each event. In addition, these reports contain detailed information on the performance of geotechnical structures, of building foundations and of infrastructure, particularly roads and bridges, dams and ports. They also contain some recommendations for reconstruction and for future building. The earthquake performance of buildings of different forms of construction is discussed in more detail in Chapter 5, and ways in which buildings can be improved are discussed in Chapter 7.

2.4 Earthquake Losses in Rich and Poor Countries

Table 2.1 lists 23 significant earthquakes which have occurred since the year 2000, causing a total of 699 000 deaths and US$488bn of damage costs. Using data on national income given by the World Bank, we can allocate these totals by the four country income categories defined by the World Bank, as shown in Table 2.2.

Table 2.2 Distribution by World Bank Income Groups of deaths and damage costs in the 23 significant earthquakes listed in Table 2.1.

World Bank Country Group	GNI/capita (2018 US$)	Deaths (% of total)	Damage costs (% of total)
Low-income	<996	33	3
Lower-middle income	996–3895	13	2
Upper-middle income	3896–12055	51	25
High-income	>12055	3	71

The income group of each country is shown in Table 2.1 (World Bank 2019).
Source: World Bank (2019).

Of the deaths, 51% have occurred in upper-middle income countries, 13% in lower-middle income countries, 33% in low-income countries and only 3% in high-income countries (principally the 2011 tsunami deaths in Japan). For damage costs, the order is reversed: 71% of the damage costs were borne by the high-income countries (mostly by Japan), a further 25% by the upper-middle income countries and only 2 and 3%, respectively, in the lower-middle and low income countries. So clearly, a country's vulnerability depends to an extent on its relative wealth. But most of the deaths from these events were in the upper-middle income countries, not the poorest. And damage costs were very significantly concentrated in the high-income countries.

Another way to look at this disparity is to plot damage costs against casualty figures country by country, using all of the earthquakes which are reported for each country in the EM-DAT database for the period 2000–2019 (CRED 2020). This plot is shown in Figure 2.19.

Dotted lines on the chart show the loss per fatality from the earthquakes during this 20-year period. The position of different countries on this chart shows both their relative earthquake-proneness and their relative wealth. Countries with the highest life loss are on the right of the chart, but those with the greatest damage cost at the top. The most vulnerable countries are those towards the bottom right (India, Pakistan, Nepal, Iran, Indonesia and Haiti). The least vulnerable to loss of life are at the top left (Chile, Italy and Mexico). Countries such as the United States and Greece in which fewer than 200 lives were lost in that period of time as a result of earthquake ground shaking do not appear on the chart at all.

2.5 Are Earthquake Losses Decreasing Over Time?

In order to investigate whether earthquake losses are decreasing over time, we need to look at both human casualties and damage costs. Both need to be viewed from a longer-term perspective than just the 20 years which has been the focus of this chapter so far. Because most of the damage and casualties occur in just a few earthquakes, it is better to assemble the loss data by decade.

Reasonably reliable data on human casualties are available from about 1900 onwards, using a variety of catalogues, and this has been assembled and maintained by our colleague

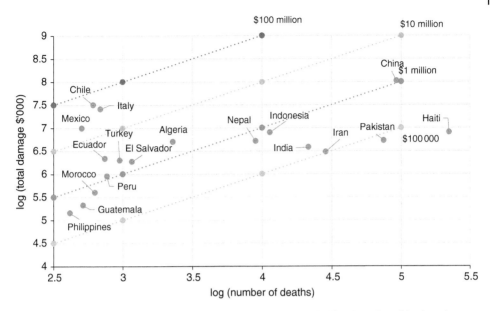

Figure 2.19 Earthquake damage and fatalities by country 2000–2019, using a logarithmic scale based on EM-DAT Disasters Database (CRED 2020). Note: this chart excludes events in which the majority of deaths were related to tsunami rather than ground shaking. *Source:* Data from CRED (2020).

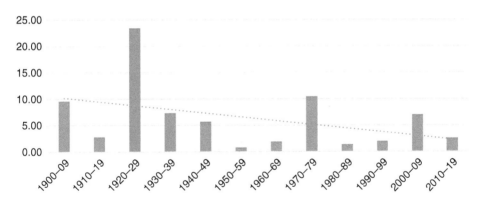

Figure 2.20 Global earthquake mortality rate since 1900 (deaths per 100000 population). *Source:* Data from Antonios Pomonis's Fatal Earthquakes database.

Antonios Pomonis (Daniell et al. 2018). The chart shown in Figure 2.20 is based on these data. Because the world's population has been growing rapidly in this period, it is important to take account of this increase in comparing the number of deaths over time. Figure 2.20 shows the earthquake mortality rate in each decade since 1900, where the mortality rate is the estimated number of deaths per 100000 of the world's population in the mid-year of the decade. Data on serious injuries are also available for many events, but injury definitions vary, so these are less useful for long-term comparisons.

Figure 2.20 shows that the mortality rate has been decreasing to some extent over the last 120 years, as indicated by the trendline. But the decline is slow and actual mortality rates vary enormously from decade to decade.

There are some reasons for thinking that this positive downward trend will continue. Fire following earthquakes is a smaller risk than it was, because of changes in building materials and fire protection. And there is also evidence that in many countries, there is increasing adherence to building codes partly as a result of greater public awareness of the earthquake risk. And specific building strengthening programmes have been undertaken in some countries (see Chapter 8). On the other hand, as Roger Musson has noted in his book *The Million Death Quake* (Musson 2012), the continuing concentration of global population into huge cities with extremely vulnerable housing is increasing the risks in many of them, and has given rise to the possibility of a single future earthquake which could cause more than a million deaths, a much higher death toll than has occurred to date in a single event. The most at-risk cities can now be identified, and efforts are being made in many of them to improve construction standards. Future trends in the mortality rate will depend crucially on what can and is being achieved in these cities. A further factor which could influence future death rates is the rapid increase of population in coastal areas (A. Pomonis, 2020, personal communication). Daniell et al. (2012) also points out that there has been, over the period 1900–2012, an increase in earthquake deaths as a percentage of total deaths, suggesting that greater success has been achieved in reducing death rates from other causes.

The costs of earthquake damage have also been assembled for over 7000 damaging earthquakes since 1900 in the CATDAT database by James Daniell and co-workers at KIT in Karlsruhe in Germany, and the trends in losses up to 2012 have been analysed (Daniell et al. 2012). Costs of earthquakes are usually divided into direct and indirect costs. The direct costs are those associated with the actual damage caused by the earthquake, while the indirect costs (which are more difficult to quantify) are those associated with loss of production and overall economic consequences of the event. To compare the losses over time, costs have been brought to a present value. The adjustment is to bring all costs to the values that would be paid in today's money for the event-year earthquake effects. The total (direct and estimated indirect) earthquake losses have then been calculated as a percentage of GWP, the Gross World Product (the world's total economic output), also adjusted to present-day values, to give an annual global loss rate.

The surprising conclusion is reached that, although the global annual loss rate has fluctuated somewhat over time, the trend has remained relatively constant, with a peak of 0.1% in 1949, reducing in more recent years to about half of this value. Thus, while production and population have increased enormously over the last century, and thus actual values of loss have increased correspondingly, there has been only a small decrease in the loss rate as a proportion of GWP. But this decrease is not enough to suggest that the problem of economic earthquake losses is being brought under control. For small poor countries such as Haiti, the total cost of a single event can be more than 100% of the country's GDP. Indeed, recent events have suggested that, given the increasing interconnectedness of the world's economy, the indirect losses are increasing and will become more dominant in future events (Daniell et al. 2012).

References

Asia Development Bank (2005). *Pakistan 2005 Earthquake: Preliminary Damage and Needs Assessment*. Islamabad: Asian Development Bank and World Bank.

Berberian, M. (2005). The 2003 Bam urban earthquake: a predictable seismotectonic pattern along the Western Margin of the Rigid Lut Block, Southeastern Iran. In: *2003 Bam Iran Earthquake Reconnaissance Report*, Earthquake Spectra Publication No 2005-04. California: Earthquake Engineering Research Institute.

Boen, T. (2016). *Learning from Earthquake Damage: Non-Engineered Construction in Indonesia*. Indonesia: Gadjah Mada University Press.

Coburn, A. and Spence, R. (2002). *Earthquake Protection*. Wiley.

CRED (2020). The international disaster database. http://www.emdat.be (accessed 2020).

Daniell, J., Khazai, B., and Wenzel, F. (2012). The worldwide economic impact of historic earthquakes. *Presented at the 15th World Conference on Earthquake Engineering*, Lisbon, Portugal.

Daniell, J., Pomonis, A., Tsang, H.-H. et al. (2018). The top 100 fatal earthquakes examining fatality risk reduction with respect to seismic code implementation. *Proceedings: Presented at the 16th European Conference on Earthquake Engineering*, Thessaloniki, Greece.

DesRoches, R., Comerio, M., Eberhard, M. et al. (2011). Overview of the 2010 Haiti earthquake. *Earthquake Spectra* 27: S1–S21.

EEFIT (2005). *The Bhuj, India Earthquake of 26th January 2001*. UK: EEFIT, Institution of Structural Engineers.

EEFIT (2006). *The Indian Ocean Tsunami of 26.12.04: Mission Findings in Sri Lanka and Thailand* (ed. A. Pomonis). London: Institution of Structural Engineers.

EEFIT (2008a). *The Kashmir Pakistan Earthquake of 8th October 2005: A Field Report by EEFIT*. London: Institution of Structural Engineers.

EEFIT (2008b). *The Wenchuan Earthquake of 12th May 2008: A Preliminary Field Report by EEFIT*. London: Institution of Structural Engineers.

EEFIT (2010). *The Haiti Earthquake of 12th January 2010: A Field Report by EEFIT*. London: Institution of Structural Engineers.

EEFIT (2011a). *The Mw 6.3 Christchurch New Zealand Earthquake of 22nd February 2011: A Field Report by EEFIT*. London: Institution of Structural Engineers.

EEFIT (2011b). *The Mw 9.0 Tohoku Earthquake and Tsunami of 11th March 2011: A Field Report by EEFIT*. London: Institution of Structural Engineers.

EEFIT (2015). *The Mw 7.8 Gorkha, Nepal Earthquake of 25 April 2015: A Field Report by EEFIT*. London: Institution of Structural Engineers.

EERI (1986). *Reducing Earthquake Hazards: Lessons Learnt from Earthquakes*. California: Earthquake Engineering Research Institute.

EERI (2005). *First Report on the Kashmir Earthquake of October 8, 2005*. EERI Newsletter 39:12. California: Earthquake Engineering Research Institute.

EERI (2006a). *The Kashmir Earthquake of October 8, 2005: Impacts in Pakistan*. EERI Newsletter 40:2. California: Earthquake Engineering Research Institute.

EERI (2006b). *The Mw 6.3 Java, Indonesia, Earthquake of May 27, 2006*. California: Earthquake Engineering Research Institute.

EERI (2008). *The Wenchuan, Sichuan Province, China, Earthquake of May 12, 2008*. California: Earthquake Engineering Research Institute.

EERI (2016). *The M7.8 Gorkha Nepal Earthquake and Its Aftershocks*. California: Earthquake Engineering Research Institute.

Foulser-Piggott, R., Spence, R., Eguchi, R., and King, A. (2015). Using remote sensing for building damage assessment: GEOCAN study and validation for 2011 Christchurch earthquake. *Earthquake Spectra* 32: 1–21.

Galloway, B., Hare, J., Brunsden, D. et al. (2014). Lessons from the post-earthquake evaluation of damaged buildings in Christchurch. *Earthquake Spectra* 30: 451–474.

Ghafory-Ashtiany, M. and Mousavi, R. (2005). History geography and economy of Bam. *Earthquake Spectra* 21 (S1): S3–S11.

Ghosh, S., Huyk, C., Greene, M. et al. (2011). Crowdsourcing for rapid damage assessment: the global earth observation catastrophe assessment network. *Earthquake Spectra* 27 (1): S179–S198.

Government of Nepal (2015). *Nepal Earthquake 2015: Post Disaster Needs Assessment, Volume A: Key Findings*. Kathmandu: National Planning Commission, Government of Nepal.

Jain, S.K. (2016). Earthquake safety in India: achievements, challenges and opportunities. *Bulletin of Earthquake Engineering* 14: 1337–1436.

Jain, S.K., Lettis, W., Murty, C., and Bardet, J. (2002). *Bhuj India Earthquake of January 26th 2001: Reconnaissance Report*. EERI Special Report. California: EERI.

King, A., Middleton, D., Brown, C. et al. (2014). Insurance: its role in recovery from the 2010–2011 Canterbury earthquake sequence. *Earthquake Spectra* 30: 475–492.

Langenbach, R. (2015). What we learn from vernacular construction. In: *Nonconventional and Vernacular Construction Materials: Characterisation, Properties and Applications* (eds. K.A. Harries and B. Sharma). Woodhead Publishing.

Macabuag, J., Raby, A., Pomonis, A. et al. (2018). Tsunami design procedures for engineered buildings: a critical review. *Proceedings of the Institution of Civil Engineers – Civil Engineering* 171: 166–178.

Maheri, M., Naeim, F., and Mehrain, M. (2005). Performance of adobe buildings in the 2003 Bam earthquake. In: *2003 Bam Iran Earthquake Reconnaissance Report*, Earthquake Spectra Publication No 2005-04. California: Earthquake Engineering Research Institute.

Marshall, J., Lang, A., Baldridge, S., and Popp, D. (2011). Recipe for disaster: construction methods, materials and building performance in the January 2010 Haiti earthquake. *Earthquake Spectra* 27: S323–S344.

Moon, L., Dizhur, D., Senaldi, I. et al. (2014). The demise of the URM building stock in Christchurch during the 2010–2011 Canterbury earthquake sequence. *Earthquake Spectra* 30: 253–276.

Movahedi, H. (2005). Search, rescue and care of the injured following the 2003 Bam Iran earthquake. *Earthquake Spectra* 21 (S1): S475–S485.

Murty, C. (2005). *Earthquake Rebuilding in Gujarat, India*. EERI Recovery Reconnaissance Report. California: Earthquake Engineering Research Institute.

Musson, R. (2012). *The Million Death Quake*. Palgrave Macmillan.

National Police Agency (NPA) (2011). Damage situation and police countermeasure, 8 November 2011 [online]. http://www.npa.go.jp/archive/keibi/biki/higaijokyo_e.pdf (accessed 9 November 2011).

Saito, K., Spence, R., Going, C., and Markus, M. (2004). Using high-resolution satellite images for post-earthquake damage assessment: a study following the Bhuj earthquake of 26.1.01. *Earthquake Spectra* 20: 145–270.

So, E. (2009). The assessment of casualties for earthquake loss estimation. PhD thesis. Cambridge University.

Spence, R. (2014). The full-scale laboratory: the practice of post-earthquake reconnaissance missions and their contribution to earthquake engineering. In: *Perspectives on European Earthquake Engineering and Seismology* (ed. A. Ansel). The Netherlands: Springer.

Spence, R., Palmer, J., and Potangaroa, R. (2009). Eyewitness reports of the 2004 Indian Ocean tsunami from Sri Lanka, Thailand and Indonesia. In: *The 1755 Lisbon Earthquake Revisited* (eds. L. Mendez-Victor, C. Oliveira, J. Azevedo and A. Ribeiro). Springer.

Tapponier, P., King, G., and Bollinger, L. (2006). Active faulting and seismic hazard in the Western Himalayan Syntaxis, proceedings, international conference on the 8.10.2005 Kashmir earthquake. *Presented at the International Conference on the 8.10.2005 Kashmir Earthquake*, Geological Survey of Pakistan, Islamabad.

WHO (2006). *Injuries and Disability: Priorities and Management for Populations Affected by the Earthquake and Tsunami in Asia*. World Health Organisation.

Wong, E. (2008). Chinese stifle grieving parents' protest of Shoddy School construction. *New York Times* (June). https://www.nytimes.com/2008/06/04/world/asia/04china.html?em&ex =1212638400&en=d068b0b604044615&ei=5087%0A.

World Bank (2019). World development indicators. http://www.worldbank.org/world-development-indicators (accessed 2019).

Yin, Y., Wang, F., and Sun, P. (2009). Landslide hazards triggered by the 2008 Wenchuan earthquake, Sichuan, China. *Landslides* 6 (2): 139–152.

3

How are Buildings Constructed in Earthquake Zones?

3.1   Introduction

A building must meet at least three requirements. It must provide spaces suitable for the activities which will take place in it. It must protect the occupants from the weather, from excessive sun, rain, snow and wind, and create comfortable living or working conditions indoors. It must be strong enough to resist the largest loads which it can be expected to experience during its lifetime.

These requirements apply wherever it is built, and whatever it is used for. They are as relevant for buildings built by their owners as they are for modern structures designed by professionals. And they apply as much for buildings to be used as dwellings as to those intended as workplaces, schools, hospitals or places of worship.

There is a great variety of ways in which these requirements can be met, in places with different climates, different available materials and skills for building, in urban and rural environments, in wealthier or poorer economies. And this has given rise to an astonishing and delightful variety of traditional built forms around the world (Duly 1979; Oliver 1987). The term *vernacular architecture* is used for these local traditions of building.

In this book, we are concerned with the response of buildings to earthquakes, just one of the impacts which they may be subjected to, and we need to understand how materials and form of construction affect earthquake resistance. The materials a building is built with are important, especially the materials of the vertical load-bearing structure, because it is through this structure that the forces exerted by an earthquake on a building will be transmitted to the ground. The weight of the building depends also on the materials of construction. Heavier buildings put larger forces into the load-bearing structure. But aspects of the form of a building are also important. Dowrick (1987) states that to have the best chance of resisting earthquakes a building should, ideally, be simple, be symmetrical, not be too elongated in plan or elevation, and have a uniform and continuous distribution of strength. These requirements may not be very specific, but they have the merit that they apply wherever earthquakes may occur.

The appropriate response of a building to the climate is also influenced by its materials of construction and its form, but is local, and differs according to the climatic zone. In some areas, for instance, climatic response favours heavyweight construction. In other areas, it

Why Do Buildings Collapse in Earthquakes?: Building for Safety in Seismic Areas,
First Edition. Robin Spence and Emily So.
© 2021 John Wiley & Sons Ltd. Published 2021 by John Wiley & Sons Ltd.

favours lightweight construction. Therefore, climatic requirements may well conflict with those for optimum earthquake resistance. And, since the climate is relatively constant (its changes are rather slow), it is likely to have a much more dominant influence on built form than resistance to earthquakes, which are generally rare events.

This chapter starts by looking at the ways in which built form is influenced by climate in different earthquake-prone areas of the world, and discusses the variety of forms of vernacular architecture which have emerged. The mismatches between these forms and the need for earthquake resistance are identified.

However, these vernacular forms of building are generally derived from rural living, with the availability of natural materials, timber, soils, and the accompanying craft skills, and the space to build. They are less suitable for urban living. Built forms in urban areas have to respond to dense living, and lack of time for material-gathering and for regular building maintenance. Thus, modern manufactured materials, multistorey construction and profes-sionalisation of construction, and the use of mechanical heating and cooling methods have become the norm. These urban building types too create problems for earthquake resist-ance, which will be discussed. The aspiration to urban living means that urban building methods also have an impact on the way that new buildings are built in rural areas.

The chapter will briefly review the current methods of building for residential buildings in the different zones of highest earthquake risk across the world, with the aim of demon-strating the wide disparities which exist globally between different countries and seismic regions.

3.2   Built Form, Climate and Earthquake Resistance

Climate has a powerful influence on built form, because indoor comfort in regions with the widely different climates across the world can be affected either positively or nega-tively by the materials of construction used, as well as by the design of the building envelope, its plan, number of stories, roof shape, openings, building orientation and building layout.

Climate is never the sole determinant of built form, because many other influences, both physical and cultural, affect building design. And as material resources permit the use of mechanical energy systems, these are increasingly used for climate control, even where they can easily be avoided by alternative design approaches.

Nevertheless, traditional building methods and especially rural building methods are often to a considerable extent a response to the prevailing climate. It is important to understand how the forms of building suggested by the climate respond in earth-quakes, and in particular to identify where there is an intersection between the form appropriate to the climate and the need for earthquake resistance, and where there are mismatches.

The world's climates can be divided into five broad climate types, each of which has many subtypes. The most commonly used climate classification system is the Köppen–Geiger classification system (Kottek et al. 2006). The five primary classes, equatorial, arid, warm temperate, continental and polar are shown in Table 3.1, along with the broad climatic criteria which define each class.

Table 3.1 The five primary classes of climatic zones in the Köppen–Geiger classification system, and the general criteria of classification.

Zone	Main climate type	Criterion
A	Equatorial	Mean temperature of every month is $>+18\,°C$ with significant rainfall
B	Arid	Low annual rainfall
C	Warm temperate	Mean temperature of the coldest month is between -3 and $+18\,°C$
D	Continental	Mean temperature of the warmest month is $>10\,°C$ and mean temperature of coldest month is below $-3\,°C$
E	Polar	Mean temperature of warmest month is $<+10\,°C$

Figure 3.1 shows the distribution of climatic zones based on past observed historic climate data (1960–2016). The boundaries of these zones are not fixed or static: significant shifts in the boundaries are expected to occur in the future as a result of climate change (Rubel and Kottek 2010).

Since most of the world's damaging earthquakes occur in zones A, B and C, the relation between climatic responsiveness and earthquake resistance will be considered for those zones.

3.2.1 Buildings for (Equatorial) Warm Humid Climates (Zone A)

The important characteristics of the warm humid climates are (Koenigsberger et al. 1973): moderately high daytime air temperatures with little fluctuation between day and night, and little seasonal fluctuation; high humidity at all times of year; high rainfall, often concentrated into two monsoon seasons, with continual dampness of atmosphere, and generally low wind velocities; and high solar radiation modified by frequent cloud cover and vegetation, resulting in little reradiation from the ground.

In these climates, the tasks of shelter are to provide protection from solar radiation by insulation and shading, and also to encourage air movement to cool the body, while providing protection from rainfall. To achieve this, buildings need to be well-spaced and openings large to allow air movement. Steeply pitched well-insulated roofs provide thermal insulation and allow water run-off. Thin and lightweight walls and roofs enable the building to cool off at night. Large eaves overhangs provide shading for the walls and protect them from erosion. Buildings are often raised above the ground. Materials traditionally used in hot humid climates have been timber and bamboo frames for walls, floors and roofs, and thatch or lightweight sheet or tiles for roofing and lightweight wall infills (Figure 3.2).

3.2.2 Buildings for Arid Climates (Zone B)

The important characteristics of arid climates are: high daytime temperatures and a large diurnal temperature range, frequently with large seasonal temperature variations; low humidity; little rainfall, usually concentrated into one short and sometimes unreliable season, and high solar radiation, with very little cloud cover, and little vegetation to reduce

Main climates

A: Equatorial   C: Warm temperature

B: Arid

D: Snow   E: Polar

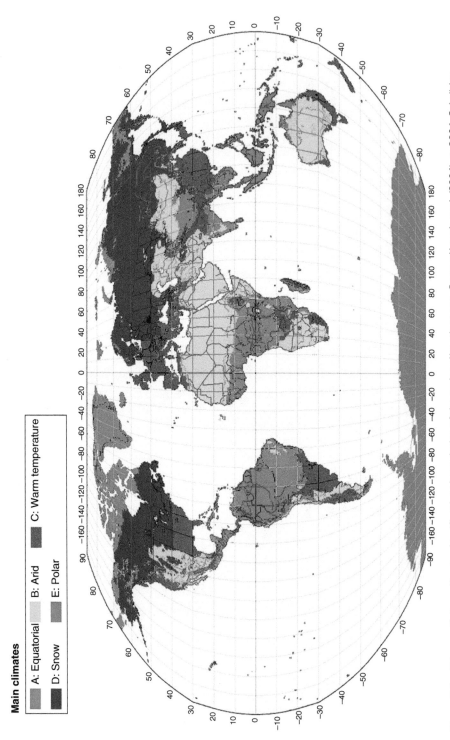

Figure 3.1 Köppen–Geiger climate classification: boundaries of the major climatic zones. *Source:* Kottek et al. (2006). © 2006, Gebrüder Borntraeger.

Figure 3.2 Timber-framed house with lightweight walls and roof in warm humid climate zone of Aceh, Indonesia, 2006. *Source:* Photo: T. Boen.

solar glare and reradiation from the ground. Local winds can also occur, creating a problem of dust- or sand-storms.

In these climates, the primary function of shelter is to modify the extremes of air temperature, and to protect the body from solar radiation and glare. Air movement is of no value for cooling, so can be eliminated, and dust needs to be excluded. To meet these requirements, buildings need to have a high thermal mass, which implies thick walls and roofs. Buildings are closely spaced to provide mutual shading, protect from winds and to reduce solar glare. Roofs are usually flat and can be used for sleeping outside in the hottest months, or for storage, Figure 3.3. Wall openings are small and set high in the walls. The thick walls required are usually provided by rammed earth or rubble stone in a mud mortar. Roofs may be of rammed earth supported on a layer of poles and brushwood, or are sometimes vaulted or domed, which altogether eliminates the need for timber (Spence and Cook 1983), which is usually scarce in these regions.

3.2.3 Buildings for Warm Temperate Climates (Zone C)

The general characteristics of temperate climates are that they are distinctly seasonal, with colder winters and warmer summers. The lowest winter temperatures are below freezing for a small proportion of the year, and the warmest summer temperatures are rarely above 35 °C. Rainfall is seasonal, but rarely heavy, and winds are variable.

Buildings responding to these requirements should generally have light, well-insulated roofs, and well-insulated walls, medium-sized openings (25–40% of wall area) positioned at

Figure 3.3 Village housing in the hot-dry area of Iran. Open courtyards, closely spaced buildings, thick, heavyweight stone walls, flat roofs and shallow domes provide for comfort, and use locally available building materials. *Source:* Paul Oliver. Reproduced with permission of Paul Oliver Vernacular Architecture Library, Oxford Brookes University.

body height to allow for light and ventilation, and open spacing between buildings; but many local variants exist.

Masonry walls and pitched tile or slate roofs are appropriate materials in many locations, but timber-framed and timber-clad buildings are also suitable. Two-storey houses are common.

3.2.4 Mismatches Between Climatic Response and Seismic Resistance

It is clear from the previous sections that the appropriate climatic response is not always consistent with an appropriate earthquake response. Where buildings are light in weight, and made of timber or other flexible materials, they are less likely to collapse under earthquake ground-shaking than buildings made of heavy and brittle materials. Likewise, the spacing and shape of buildings and the size and position of openings can affect both their climatic response and their earthquake response.

A first attempt to quantify these mismatches was made by W. Victoria Lee (2008). Using climatic design data (Koenigsberger et al. 1973), Lee has proposed a seismic responsiveness score for the seven building characteristics which would represent an optimum climatic solution for cities in each of 26 of the world's climatic zones and subzones. The result is shown in Figure 3.4. In this chart, the highest possible seismic responsiveness score is 14. Buildings built to be responsive to the climate in Zones A and C perform relatively well, while those in the arid zones (Zone B) perform poorly and those at high elevations (greater than 1000 m) are particularly poor. This is because of the need for heavyweight materials in those zones to provide heat storage. It is made worse in areas where brittle mud or stone must be used since these can be lethal when they collapse on occupants. Some of the world's most lethal earthquakes such as that in Bam, Iran in 2003 and Ancash, Peru in 1970 have occurred in these arid zones.

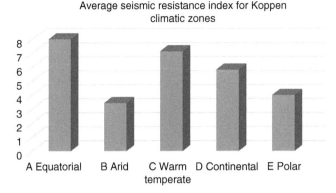

Figure 3.4 Estimated seismic resistance index for buildings designed with appropriate response to each of the world's major climate zones (A = equatorial, B = arid, C = warm temperate, D = snow, E = polar; seismic index 0–4 = poor, 4–8 = moderate, 8–10 = good). *Source:* Modified from Lee (2008).

3.3 Building Construction Types by Earthquake Zone

3.3.1 Introduction

The seismicity of the continents can be seen at a glance in the 2018 GEM Global Seismic Hazard map (http://www.globalquakemodel.org/gem), a picture of the world's seismicity put together in 2018 by combining the efforts of more than 30 local and regional groups (Figure 3.5).

The map shows, by colour contouring, the level of ground-shaking, measured as the peak ground acceleration (PGA) with a 10% probability of being exceeded once in 50 years (the 475-year return period), at each point on the earth's land surface. The areas shown in orange and red have the highest expected ground-shaking, greater than 0.55 g, a level which would result in devastation for buildings not built to withstand an earthquake. These areas are a relatively small part of the earth's surface, but they are also the areas of greatest importance for earthquake safety. Adjacent to them is a much larger area shown in yellow, where the expected 475-year return period ground-shaking would be greater than 0.2 g. At this level, ground-shaking would be seriously damaging to buildings not specifically designed for earthquakes, and potentially destructive to particularly vulnerable building types. And even in the areas (shown in green and light yellow in Figure 3.5) where the 475-year return period shaking level is between 0.1 and 0.2 g, vulnerable building types could still be seriously damaged.

The map of Figure 3.5 shows three broad regions of the highest seismicity. The first includes much of the Pacific coastal region of South, Central and North America. The second region includes the Pacific margin of East and South East Asia, including Japan, Taiwan, most of South East Asia and also New Zealand. The third region is known as the Alpine-Himalayan zone which extends from Southern Europe, through Turkey and Iran to Pakistan, Nepal, northern India and south-west China. There are in addition, some areas of significant seismicity away from these highest risk zones, where the 475-year return period ground-shaking is 0.1 g or more, including parts of Eastern North America, parts of North

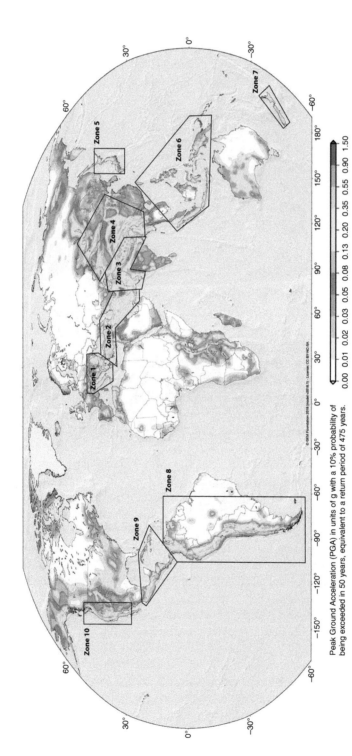

Figure 3.5 The GEM global seismic hazard map with the 10 zones identified in this chapter. *Source:* Global Earthquake Risk Map, www. globalquakemodel.org/gem. © Global earthquake maps.

Peak Ground Acceleration (PGA) in units of g with a 10% probability of being exceeded in 50 years, equivalent to a return period of 475 years.

0.00 0.01 0.02 0.03 0.05 0.08 0.13 0.20 0.35 0.55 0.90 1.50

Africa and the Rift Valley in East Africa, substantial parts of China and Eastern Russia, and the Pacific Islands.

The high seismicity zones vary widely in population density and living standards, and encompass many different climatic zones. For all these reasons there are wide variations in the types of buildings people have built and are building today. To examine the buildings in these areas more closely, taking account of some of these differences, we have subdivided the high seismic hazard areas into 10 seismic zones, and assembled available building data by country groups or earthquake zones. Our use of data from the countries within these zones depends on their size and what we know about their current residential building stock.

The first four zones are all part of the Alpine-Himalayan belt. Zone 1 (Southern Europe) includes Italy, Greece and Romania. Zone 2 (West Asia) includes Turkey and Iran. Zone 3 (South Asia) includes Pakistan, Nepal and India. Zone 4 is south-west China. The next three zones are on the western Pacific margins: Zone 5 is Japan, Zone 6 is the South-East Asian countries of Indonesia and the Philippines and also Taiwan, and Zone 7 is New Zealand. The final three zones are located on the western Pacific margins: Zone 8 includes the Andean countries of South America (Chile, Ecuador, Peru and Colombia), Zone 9 is Central America (El Salvador, Guatemala and Mexico) and the Caribbean; and Zone 10 is the US State of California. The building stock data available for these countries and their regions of highest seismicity are by no means complete; some of the data is not up-to-date, and it is usually available only for residential buildings. But by distinguishing these 10 seismic zones, we are able to make a broad comparison of the characteristics of the residential buildings to be found there. And this reveals some startling differences between the zones.

One source of building stock data which we have used to make this comparison is the United States Geological Survey (USGS) Global Building Inventory (Jaiswal and Wald 2014), which was created to inform early post-earthquake assessments of the damage and casualties likely to have resulted, to assist in relief and aid efforts, through its PAGER programme (Jaiswal et al. 2011). The USGS inventory provides four country by country tables, for urban and for rural buildings, each divided into residential and non-residential. For each of these classes, the total number of buildings is not given, but the proportion falling into each of 106 different construction types is estimated. The types are defined by the form of construction, primarily the structural load resisting system used (wood, steel, reinforced concrete or masonry) with subdivisions according to type of vertical elements, materials and other important characteristics affecting earthquake performance. There are very few countries for which this kind of data is available in full, and to achieve a complete inventory, a good deal of local expert judgement has been used in the USGS study. The initial study was done in 2008, and has been updated in 2014 with a plan to update it further in future. Inevitably the relative occurrence of different construction types will vary with time, as new buildings are added, and some older ones are demolished. For several of the countries, additional local sources of data have been used.

Using the USGS PAGER inventory only, we show, in Figure 3.6, the distribution in each of our 10 seismic zones of the most numerous class, the urban residential buildings, between the four major construction types (timber, steel, reinforced concrete and masonry) and a further class, 'other' which includes other materials and cases where the type is unknown. The summation across all the countries in each zone has taken account of the 2017 relative urban populations of each country (https://data.worldbank.org).

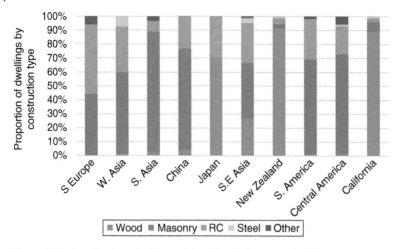

Figure 3.6   Distribution of urban residential building types by 10 seismic regions. Building stock data from USGS (*Source:* Jaiswal and Wald 2014); urban population data from World Bank (https://data.worldbank.org).

The comparison between the 10 seismic regions reveals some interesting and important differences in the building styles between these regions. In three of them, North America, Japan and New Zealand, wood-frame construction predominates. This is not necessarily because the climate or the building materials available make this form of construction preferable. It is mainly because wood is an easily accessible and affordable local material; to some extent, at least in Japan and New Zealand, it is a response to the experience of damage to other types of building (notably masonry) in past earthquakes, which has today become codified in acceptable construction practice.

In Central and South America, and also in West Asia, South Asia and China, masonry buildings predominate. However, very different styles of masonry are found in these five different regions, some of them built with seismic safety in mind (confined masonry), and others using masonry techniques which have a very poor performance in earthquakes. These differences are described and discussed in the following sections.

Finally, in Southern Europe, we find that the predominant construction type in urban areas is reinforced concrete, with a significant minority of masonry buildings. Reinforced concrete is often in the form of multistorey apartment buildings, which may or may not have good earthquake resistance, depending on the date and quality of their construction. Much of the masonry in the region, particularly traditional stone masonry, has a relatively poor performance in earthquakes.

The following sections look in more detail at what can be deduced from the available data on the building methods adopted in the different regions. A number of local reports are used to describe typical construction methods. An important source of information for this discussion is the World Housing Encyclopedia (WHE) (http://www.world-housing. net). WHE is a collection of resources related to housing construction practices in the seis-mically active areas of the world. Its aim is to share experiences with different construction types and encourage the use of earthquake-resistant technologies worldwide. WHE con-tains over 140 housing reports from more than 40 countries, written by local architects or engineers. Each gives details of construction techniques, where built and by what process, and information on performance in past earthquakes.

3.3.2 Zone 1: Southern Europe (Italy, Greece and Romania)

In all of these countries, the highest risk countries in Southern Europe, there is a high level of urbanisation, with urban populations of 79% in Greece, 74% in Italy and 54% in Romania. But many of the cities are old-established, and have many buildings more than 100 years old, so modern construction materials are not as dominant as in cities elsewhere. Urban residential construction is a mixture of reinforced concrete and masonry construction, and methods of building are a reflection of the age of construction, the economic and political regime, the development of codes of practice and the degree of municipal control over building quality.

In all these countries, 5- to 11-storey apartment buildings in reinforced concrete are a dominant feature of urban building in recent decades. Buildings built before the 1950s (depending on when the earthquake codes were introduced) will have reinforced concrete frames with masonry infills, but without specific design for earthquake loading. More recent buildings will have reinforced concrete shear walls or frame elements designed for lateral loads (Tassios and Syrmakezis 2002) (Figure 3.7), although some illegally built structures without proper earthquake-resistant design are still being built.

Figure 3.7 Reinforced concrete dual frame and shear-wall apartment building in Greece. *Source:* Antonios Pomonis. Reproduced with permission.

A feature of many apartment blocks is the use of the ground floor for commercial space or car parking. The open plan and extra height of these floors leads to a so-called 'soft storey', which has been a source of weakness in many earthquakes.

In Romania, during the era of State planning, large numbers of apartment buildings were built using precast concrete panel systems, which have good record of performance in earthquakes.

Masonry buildings, usually two to five storeys high, constitute a very important part of the urban and rural building stock in all the European countries. The oldest buildings may have walls built in stone (even earthen construction in Romania), with brittle and poorly bonded masonry and timber or vaulted floors (D'Ayala and Speranza 2002) (Figure 3.8). These buildings are highly vulnerable to earthquakes, and have been responsible for numerous earthquake disasters (notably in Italy). Efforts have been made to strengthen such buildings in recent years, with limited success, and replacing them is difficult, as they often form the valued historical core of the town or city. More recent masonry is built in brick or hollow clay block, with reinforced concrete floor slabs or confined masonry, as described in Chapter 5, and these buildings have performed better in earthquakes.

Figure 3.8 Stone masonry urban houses in Italy. *Source:* Randolph Langenbach. Reproduced with permission.

3.3.3 Zone 2: West Asia (Turkey and Iran)

Turkey and Iran are the most earthquake-prone countries in this part of the Alpine-Himalayan seismic region. They are similar in population (about 80 million) and level of urbanisation (about 81% in 2017), but differ in average level of income (in 2018, Turkey had approximately twice that of Iran) and climate. Both have a history of devastating earthquakes in recent years, so there is considerable public awareness of the earthquake risk, and incentive to government and building owners to improve standards of earthquake resistance.

In both countries also, rural building is largely of masonry, with traditional stone and adobe construction dominating. With its arid climate, adobe was traditionally more commonly used in southern Iran (Mehrain and Naeim 2004), (Figure 3.9), and stone was more commonly used in Turkey and Northern Iran. More recent rural construction has used unreinforced clay masonry (bricks or blocks) in both countries. Rural buildings have performed very poorly in repeated earthquakes in recent decades. In the 2003 earthquake in Bam, Iran, the death of over 40 000 people, mostly urban residents, was attributed primarily to the collapse of adobe masonry buildings.

Urban construction differs between the two countries. In Turkey, the majority (60%) of urban construction is in the form of apartment blocks in reinforced concrete. These tend to have between five and seven storeys, and use a frame system, often with an open ground floor, creating a soft storey (Gülkan et al. 2002) (Figure 3.10). Such buildings performed very poorly in the Mw7.6 Kocaeli earthquake in 1999 and the Mw7.1 Van earthquake in

Figure 3.9   Rural adobe house in Iran. *Source:* EEFIT. Reproduced with permission.

Figure 3.10 Reinforced concrete frame apartment building in Turkey. *Source:* EEFIT. Reproduced with permission.

2011 resulting in many collapses and deaths. Poor building control is often cited as the main reason for this poor performance. Since 1999, construction standards have improved in Turkey and new more earthquake-resistant techniques, using shear walls, and also tunnel-form construction, have been extensively used for mid- and high-rise construction (Yakut and Gülkan 2003). In Iran, only 5% of urban residents live in reinforced concrete buildings, but 16% live in multistorey steel-frame buildings.

In both countries, masonry construction is used for a very significant part of the urban residential building stock. Some are in unreinforced brick and block with reinforced concrete roofs, but confined masonry is also used. However, rapid urbanisation has resulted in a significant number of urban dwellers living in informal settlements where construction standards are very poor, and these are highly vulnerable to earthquakes.

3.3.4 Zone 3: South Asia (India, Pakistan and Nepal)

Between them, these three countries had (as of 2017) a population of about 1600 million people, more than 20% of the world's total population, and also a high proportion of those living in the world's highest earthquake-risk zones. All have experienced devastating earthquakes in the last 20 years, with very substantial loss of life. All are relatively low-income countries, and the population today is still largely rural (66% in India, 64% in Pakistan and 81% in Nepal), though urbanisation is rapid.

Throughout all three countries, low-rise masonry construction predominates, although, particularly in Nepal, timber-frame construction is used in some forested areas. But methods of masonry construction vary widely, both in the masonry materials used and the type of mortar used for bonding. According to Bothara et al. (2018), the most common forms of masonry in each country are: clay brick or stone masonry, bonded with a mud mortar (34% over the three countries); adobe blocks bonded with a mud mortar (28%); and cement-mortar bonded brick or stone (18%). Overall, masonry thus constitutes over 85% of the rural housing stock of these three countries. There are some interesting local masonry traditions in which timber framing is combined with masonry construction, such as Dhajji Dewari (Hicylmaz et al. 2011) (Figure 3.11) and these have reportedly performed much better in earthquakes than other forms of masonry construction (Langenbach 2015) (Box 3.1). But these traditional building types have not been commonly built in recent years, partly because of declining availability of timber for construction. Masonry buildings are likely to continue to predominate in rural areas, because of the availability of the material, and their affordability and simplicity of construction which suits the rural economy. Many such buildings have performed badly in recent earthquakes such as the 2001 Bhuj, 2005 Kashmir and 2015 Gorkha, Nepal events. But following these events, there have been initiatives to introduce confined masonry in each of these countries (Brzev 2020. Personal Communication).

Although a smaller proportion of the total building stock, the cities in the region have a rapidly growing number of reinforced concrete frame buildings (around 10% of the urban building stock). This building type has been tested in the same earthquakes, and many of them have not performed well (as described in Chapter 2).

Throughout the seismic zones of these countries, there have been substantial programmes to improve rural construction methods through builder-training, which will be discussed in Chapter 7.

Figure 3.11 Dhajji Dewari building in Srinagar, Kashmir. *Source:* R. Desai. Reproduced with permission.

Box 3.1   Profile: Randolph Langenbach

Randolph Langenbach: conservation of earthquake-resistant vernacular architecture

Randolph Langenbach is an architectural historian and building conservator who has pioneered the study of the earthquake resistance of buildings of vernacular architecture and made important contributions to the development of earthquake-resistant design using traditional low-cost materials.

His research on the conservation of masonry buildings in earthquake areas was first inspired by earlier work on the vast brick and stone New England textile factories which withstood the

Source: Randolph Langenbach. Reproduced with permission.

intense vibrations of the looms for over a century. After suffering the devastating setback of the loss of almost all of his prior work as a writer and photographer when his California home was destroyed in the 1991 Oakland firestorm (including the loss of over 200 000 photographs), he resumed this project with an investigation of the traditional Ottoman *himis* construction that survived the 1999 earthquakes in Turkey while neighbouring reinforced concrete structures collapsed. He has investigated and written about similar observations of the superior performance of traditional structures in the 2005 Kashmir earthquake in Srinagar, India, in Haiti after the 2010 earthquake, and more recently in Nepal after the 2015 earthquake.

His book *Don't Tear it Down* gives a detailed description of the two traditional forms of construction found in Indian Kashmir, *taq* and *Dhajji Dewari*, both using a combination of timber lacing with masonry construction; and it makes a passionate case for the preservation of these buildings and construction techniques, 'Not in spite of the antiquated construction of these buildings but because of it; (the book) is framed on the capacity of the best examples of Kashmiri construction to resist one of nature's most prodigious forces – earthquakes'.

Concerned by the frequent collapse of multistorey reinforced concrete buildings in recent earthquakes, he has been advocating the use of masonry 'armature crosswalls' in multistorey reinforced concrete buildings as a means to introduce the collapse resilience to large earthquake loads which so many reinforced concrete buildings lack. An armature crosswall is a masonry infill wall inserted into the structural frame which has the capacity to absorb energy as it deforms. Langenbach's proposals on this have been presented at international engineering conferences, though not so far implemented.

He has also had some success with the proposal to use gabion bands to create a way of bonding rubble stone masonry walls to create structural continuity within the load-bearing walls while using traditional masonry materials and skills with a minimum of imported materials. This has been applied to a number of houses and a school building reconstructed in Nepal after the 2015 Gorkha earthquake.

School building in the Nepali village of Chupar, reconstructed after the 2015 earthquake, using gabion bands. *Source:* Randolph Langenbach. Reproduced with permission.

Through his writings and lectures and his website http://www.conservationtech. com, Langenbach continues to campaign for a proper respect for the construction techniques of vernacular architecture, for the conservation rather than replacement of these buildings and for the wisdom of those who developed them as a response to the local climate, materials and occasional earthquake shocks to be respected and built upon.

Overall, construction standards in South Asia remain a very serious problem for earthquake safety, at a time when the population and particularly the urban population are increasing very rapidly. As pointed out by Sudhir Jain (2016), 'Despite having a huge problem of unsafe buildings that currently exist, we will also be constructing a huge number of new buildings in the coming years. Hence there is a tremendous urgency to ensure that all new constructions are not only equipped with basic facilities, but also should be safe'.

3.3.5 Zone 4: China

China's building stock has a particular importance, both as the world's most populous country, and because China has suffered, by some margin, more earthquake deaths in the last century than any other country. It is difficult, nevertheless, to get a clear understanding of the current building stock of China, because there is no national housing census, and because China is and has been for some years in a state of rapid change and economic development. According to World Bank data, between 1960 and 2017, China's population grew from 667 million to 1386 million, but in the same period its per capita income increased from $89 to $8826, an increase of nearly 100 fold at equivalent US$ rates, a growth rate more than 5 times that of the United States in the same period (https://data.worldbank.

org). In response to China's development policies, urbanisation has also been rapid, with a change from only 16% of the population living in urban areas in 1960 to 58% in 2017. Thus, survey data about the type and distribution of the residential building stock quickly become outdated.

Data from the USGS Global Inventory (Jaiswal and Wald 2014) indicated that in 2007 more than 90% of China's rural building stock was of masonry construction, about half of this in earthen construction (adobe or rammed earth) or rubble stone and half in unit masonry (cut stone, fired brick or concrete block). In urban areas also, masonry predominated (74%), but there was a small but growing proportion of reinforced concrete buildings. Timber-frame construction accounted for 10% of rural construction and 5% of urban construction. As in South Asia, many local traditions existed in the more remote parts of rural China, such as the Qionglong buildings of Western Sichuan and Yunnan Provinces (Figure 3.12), built either of stone or rammed earth (Wan 2013).

Both in urban and in rural areas, new building types have been appearing, in response both to the rapid urbanisation rate, and to a national New Countryside Construction initiative from the Ministry of Housing, which promotes a form of confined masonry construction.

Many recently constructed buildings were tested in the massive Mw7.9 Wenchuan earthquake of 12 May 2008, which caused 87000 fatalities, mostly from collapsed buildings. Reports from field reconnaissance missions (EEFIT 2008; EERI 2008) indicated that five- to seven-storey reinforced concrete frame buildings generally did not collapse, even where the ground-shaking was higher than buildings were designed for (Figure 3.13). Where unreinforced masonry was used for residential or for school buildings, however, there were many collapses, and numerous occupants, including school children, were killed. More

Figure 3.12 Qionglong building in Sichuan Province. *Source:* Wan Li. Reproduced with permission.

recently built confined masonry buildings were widely used in urban residential construction and performed better. The 2008 earthquake also resulted in a massive reconstruction programme in both urban and rural areas, which has brought better construction standards to the affected area of Sichuan Province, conforming to more recent codes and using some innovations in the use of traditional techniques.

3.3.6 Zone 5: Japan

Japan needs to be treated as a special seismic zone, because of the uniqueness of its residential building stock and culture. Both in urban and rural areas, wood-frame single-family houses constitute the overwhelming majority of the residential building stock. Structurally, there are two distinct types. Traditional houses, from before the 1980s, used a traditional heavy post and beam construction (shinkabe or okabe), which lacked diagonal bracing, and had a heavy tiled roof (Figure 3.15). Modern housing uses a stud-frame system, with closely spaced posts, and diagonal bracing or plywood or particle board sheathing to provide lateral resistance, and a lightweight roof on timber joists (Maki and Tanaka 2002), (Figure 3.14).

Today, more than 90% of Japan's population lives in urban areas. In 2014, in urban areas, 30% of all residences were made of timber frames of the traditional type, and 40% of the modern type; these included some multistorey apartment blocks. The other 30% of urban

Figure 3.13   Damaged reinforced concrete frame building in Dujianyang after 2008 Wenchuan earthquake.

Figure 3.14 Ordinary urban wooden housing in Japan 1970s and 1980s. *Source:* EERI. Reproduced with permission.

homes at that date were of reinforced concrete shear-wall construction. In rural areas, half of residences were traditional timber-frame structures and the rest were modern timber-frame structures with a very small proportion of concrete buildings. Newly built housing in Japan is today still very largely of timber-frame construction.

A significant proportion of the houses in Kobe in 1995 were of traditional construction, and these suffered very badly in the Mw6.9 Kobe earthquake of that year (Comerio 1998), partly because of their inherent structural weaknesses, notably the heavy roofs and a lack of horizontal bracing, but also because of their deteriorated condition. Figure 3.15 shows a drawing of the traditional shinkabe house form commonly used in Japan, Many such houses suffered complete collapse, and this contributed to the large death toll in that event. Modern timber-frame housing, by contrast, has had a better performance in recent earthquakes.

3.3.7 Zone 6: South East Asia (Indonesia and the Philippines)

Indonesia and the Philippines are the two countries we have taken to represent this part of the west Pacific seismic belt. Both countries have a large area of high seismicity and have frequently experienced damaging earthquakes; both are situated in the warm humid tropical climatic zone. They are also at comparable levels of urbanisation (47% for the Philippines, 55% for Indonesia). So a strong similarity of building construction type would be expected.

Traditionally, in both countries, wood-frame single-family houses have predominated, as would be the expected climatic response (see Section 3.1). The wood frame would be

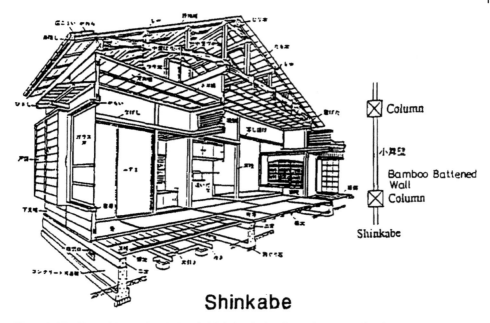

Shinkabe

Figure 3.15 Typical form of traditional shinkabe timber frame house used in Japan.
Source: EEFIT. Reproduced with permission.

infilled with thatch or mud, and have a pitched wooden roof with thatch or sometimes clay tile roofing and generous overhangs. These buildings were often built on stilts. Wood-frame buildings are still used for about 40% of rural housing in Indonesia especially in Sumatra (Figure 3.2), but less in urban areas. Because of their light weight and ductility, they have generally performed well in past earthquakes (Boen 2016; EEFIT 1990).

In the Philippines, wood-frame houses, using more modern materials for wall infills and roofing, still constitute two-thirds of the total housing stock, rural and urban. Where these have been replaced in urban areas, it is with reinforced concrete buildings (mostly non-engineered) which today constitute about one-third of the building stock. These reinforced concrete buildings are frequently very poorly constructed, and have generally performed far worse than the timber-framed buildings (EEFIT 1990). Masonry buildings are little used.

In Indonesia, resulting both from increasing spread of urban building methods, but also the unavailability of timber and the associated building skills, there has been a gradual replacement of timber-frame houses with modern materials over the last 50 years, and today, masonry buildings predominate. Masonry buildings now constitute over half of all buildings in urban and rural areas of Indonesia. A common form is a single-storey building of half-brick thick masonry, and often the masonry is confined with horizontal and vertical reinforced concrete confining elements (Boen 2016) (Figure 3.16).

In some areas, unreinforced masonry is used, of burnt clay bricks or concrete blocks in lime or cement mortar (Figure 3.17). A special feature of some volcanic areas of Indonesia is the use of lime-trass masonry blocks, in which cement is replaced by a mixture of slaked lime and a highly reactive volcanic ash (Spence and Cook 1983). In either case, timber roof beams or trusses sit directly on the walls with poor connections.

Figure 3.16 Confined masonry house in Bengkulu, Indonesia, 2007. *Source:* T. Boen. Reproduced with permission.

Figure 3.17 Traditional masonry house in Yogjakarta, Indonesia, 2006. *Source:* T. Boen. Reproduced with permission.

Confined masonry buildings have a good record of performance in earthquakes in Indonesia, but unreinforced masonry buildings have performed poorly in recent earthquakes such as that in Yogyakarta Province in 2006, Chapter 2 (So 2009).

3.3.8 Zone 7: New Zealand

New Zealand has to be treated as a separate seismic zone, because of the special development of its housing construction methods in response to repeated earthquakes. The early European settlers in New Zealand in the nineteenth century were from Europe, and tended to build in masonry as in their home countries, though some used timber-framed buildings (Beattie and Thurston 2006). Early earthquakes in 1848 and 1855, and again the devastating Mw7.4 Napier earthquake of 1931 caused severe damage to masonry buildings and also to the masonry chimneys of timber-framed buildings, so both practice and more recently (since 1944) regulations have limited the use of masonry, and provided safe detailing guidance for timber-framed buildings. Figure 3.18 shows a typical example. As a result, shaking damage to single-family residential buildings in more recent events (Mw6.5 Edgecumbe 1987 and Mw6.1 Christchurch 2011) has been largely limited to brick veneer, masonry chimney and foundation damage, and ground floor distortion ('racking') of some older

Figure 3.18 Typical New Zealand wood-frame house.

Figure 3.19 Soft-storey damage to older wood-frame house in 2011 Christchurch earthquake. *Source:* Andrew Buchanan. Reproduced with permission.

buildings (Figure 3.19), with virtually no collapsed buildings and no associated loss of life (Buchanan et al. 2011).

Timber-frame buildings, using stud-wall construction and timber truss roofs, now constitute about 90% of New Zealand's residential buildings in both urban and rural areas. A small proportion of multifamily residential buildings are built in reinforced masonry, with hollow concrete blocks and vertical reinforcement, and those in reinforced concrete use shear wall construction. Overall, the residential buildings of New Zealand are some of the most earthquake-resistant in the world.

3.3.9   Zone 8: South America (Chile, Peru, Ecuador and Colombia)

The region of high seismicity in South America passes through all the Andean countries, of which Chile, Peru, Ecuador and Colombia are chosen as the most representative. This region has been shaken, over the last century, by many destructive earthquakes, including the Mw6.1 Quindio earthquake in Colombia in 1999, the Mw8.8 Maule earthquake in Chile in 2010, and the Mw7.8 Muisne earthquake in Ecuador in 2017, all of which resulted in numerous casualties. As a result, the building stock of the zone has developed with an awareness of the earthquake risk, and standards for building have developed which are directed towards reducing the vulnerability of buildings, and applied to some degree in non-engineered housing, not just in urban areas, but in some rural areas as well.

As a result of differences in per capita income and the extent of urbanisation, earthquake experience and available materials, the four countries have rather different distributions of their residential building stock. Chile has both the highest per capita income, and also the greatest historical experience of damaging earthquakes. As in many of the highest risk zones, masonry is the material most commonly used for housing construction, and

Figure 3.20 Confined masonry apartment buildings in Chile. *Source:* Svetlana Brzev. Reproduced with permission.

accounts for two-thirds of the total housing stock. Most of this is in the form of either reinforced masonry or confined masonry (Figure 3.20), methods developed to provide for better earthquake resistance; but still about 20% is of the more vulnerable unreinforced masonry. The remaining one-third of the housing stock is of timber-frame construction. Where multistorey apartment buildings are used, these are either of confined or reinforced masonry or of reinforced concrete shear-wall construction, again with good earthquake resistance (Figure 3.21). All of these housing types have performed relatively well in recent earthquakes. In the 2010 Maule earthquake in Chile, the relatively low death toll of 562 can be attributed in part to the good performance of masonry construction.

Confined masonry is also widely used in all the other countries of the region, though there is a significant amount of unreinforced masonry using either brick or stone or even adobe (which constitutes about one-third of the housing stock in Peru). Experience has shown adobe construction to have a very poor performance in earthquakes. Figure 3.22 shows a recent estimate of the distribution of the housing stock in each country between the principal types of construction (Yepes-Estrada et al. 2017).

Because of the affordability and availability of materials and skills, masonry is likely to continue to be used, especially in rural areas of the Andean countries, and numerous projects have been devoted to developing and training and dissemination methods to improve the construction techniques using the traditional materials. Different projects have focussed on reinforced adobe in Peru (Blondet et al. 2006, 2011), improved rammed earth in Ecuador (Dudley 1993) and improved quincha (wattle and daub) in Peru (Schilderman 2004). However, these projects have mostly been implemented in the context of rebuilding after a damaging earthquake, and the improved techniques appear so far to have had rather limited application.

Figure 3.21 Concrete shear wall apartment building in Chile. *Source:* EERI. Reproduced with permission.

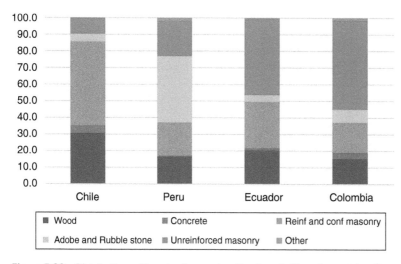

Figure 3.22 Distribution of housing by construction type in Zone 8 countries. *Source:* Yepes-Estrada et al. (2017).

3.3.10 Zone 9: Central America (Mexico, Guatemala and El Salvador) and the Caribbean

Extending northwards from Zone 8, the Pacific coast region of high seismicity passes through these three countries, along with Nicaragua and Costa Rica. Earthquakes have regularly occurred in these countries: some of them were deep events from the offshore subduction zone, while others were shallow onshore crustal events, which, though smaller in magnitude, can be more damaging to buildings. In Mexico, the Mw8 Mexico City earthquake of 1985 was of the former type, and was devastating primarily to tall buildings in Mexico City, some 330 km from the epicentre. In El Salvador, the Mw5.7 1986 and Mw7.7 2001 events, though smaller in magnitude, were shallow and affected densely populated regions. The Mw7.5 Guatemala earthquake of 1976 was a shallow onshore event, which also struck a densely populated region. In all these cases, the capital city of the country was seriously affected at a time of rapid urbanisation, and many casualties resulted.

The pattern of building in the three selected countries has been to some extent a response to the experience of these and other past earthquakes, but also differing social and economic factors. All three countries are middle-income countries, with a 2017 urbanisation level varying from 81% in Mexico to 51% in Guatemala. In the urban areas, there is a growing use of modern contractor-built methods of construction (Rodriguez and Jarque 2005) (Figure 3.23), while in the rural areas, traditional owner-built homes are still the norm. In urban areas of Mexico, reinforced concrete frame buildings now constitute more than 20% of housing stock.

Construction methods are changing in rural areas as well. Traditionally, adobe construction predominated, but adobe buildings have performed very badly in successive earthquakes, and (encouraged by the governments in each country) recent building of small homes, both in rural and urban areas, has been in confined masonry, using hollow clay or concrete blocks confined by reinforced concrete confining elements at corners and door openings and horizontal reinforced concrete tie-beams at floor levels (Jofre et al. 2011; Tena-Colunga et al. 2011). Propelled by the experience of the 1976 earthquake, confined masonry now constitutes more than 65% of housing in Guatemala (Figure 3.24), and a smaller but growing proportion in Mexico and El Salvador, while adobe construction still constitutes over 40% of rural residential construction in Mexico.

Figure 3.23 Reinforced concrete frame construction in Mexico. *Source:* EERI. Reproduced with permission.

Figure 3.24 Confined masonry housing in Guatemala 2019.

A traditional form of construction practised in El Salvador is bahareque. Bahareque, which has been used for many centuries, is a form of timber-frame construction, in which timber posts are embedded into the ground and horizontal timber members are tied to these posts at regularly spaced intervals. Smaller timber branches are then woven into these, forming a basketwork skeleton, and mud is then plastered on both sides forming a solid wall. Well-built bahareque buildings have been found to perform well in earthquakes (Lopez et al. 2004), but over time untreated timber rots and decays, so older buildings have regularly collapsed although with relatively little loss of life. Thus the use of bahareque, like adobe, has declined over the years, and now constitutes less than 10% of rural building stock in Salvador. Yet, it has been argued (Lopez et al. 2004) that this is the construction practice which best suits rural Salvador, being affordable and easy to build with local materials. As also with the similar quincha construction in Peru, efforts have been made to promote the use of improved bahareque for rural housing (Figure 3.25).

3.3.11 Zone 10: California

The final zone of high seismicity we will consider is on the Pacific coast of the United States, specifically in the state of California. The well-known San Andreas Fault and a number of associated faults pass through this region, and these faults have generated numerous highly damaging earthquakes, including the M7.8 San Francisco earthquake of 1906, the Mw6.4 Long Beach earthquake in 1933, the Mw6.9 Loma Prieta earthquake in 1989 and the Mw6.7 Northridge earthquake in 1994.

California is one of the wealthiest regions in the world, and has both the financial resources and the level of local government organisation to be able to tackle the earthquake problem. For more than a century, wood frame has been the dominant form of construction for housing, and now constitutes more than 98% of all single-family housing both in urban

Figure 3.25   Bahareque reconstruction after the 1986 earthquake in El Salvador. *Source:* M. Lopez. Reproduced with permission.

and rural areas (Arnold 2002). Today's typical form of construction, using stud-frame walls to which plywood sheathing is attached to transmit the lateral forces, and bolted connections between concrete foundation beams, walls, floors and roof, ensures a clear and continuous load path from roof to foundations (Figures 3.26 and 3.27). Building regulations, which have developed in response to observations of damage in past earthquakes, define the detailing required to resist the design earthquake in that location (Arnold 2002), making structural analysis unnecessary. As a result of these regulations and their enforcement, damage to residential buildings in the earthquakes of the past 30 years, though extensive, has resulted in few collapses or casualties.

There have nevertheless been some failures of wood-frame housing in past Californian earthquakes. In the 1989 Loma Prieta earthquake, centred near Santa Cruz, chimneys and porches collapsed, and some homes slipped off their foundations, but most of the damage to wood-frame structures was cosmetic (Comerio 1998). In the 1994 Northridge earthquake, located in the San Fernando Valley north of Los Angeles, single-family housing again performed well, but a number of multistorey apartment blocks, especially those with open ground floors for 'tuck-under' car parking suffered serious damage or collapsed with some loss of life (Comerio 1998; EERI 1994).

A small proportion of the housing in large cities, especially in the Los Angeles and San Francisco regions, was for some years in the form of multistorey brick masonry construction, but following poor performance of such buildings in earthquakes, regulations were put in place in the 1970s and 1980s requiring such buildings to be strengthened or demolished within a limited time period (Alesch and Petak 1986). Nearly all those which now remain have been preserved and retrofitted because of their heritage value.

Figure 3.26 Typical wood-frame single-family housing in California. *Source:* EERI. Reproduced with permission.

Figure 3.27 Wall framing system used in wood-frame housing construction in California. *Source:* EERI. Reproduced with permission.

3.4 Summary

Climate and the available local building materials have traditionally had an important effect on the way buildings have been built. Local vernacular building traditions have been in part a response to climatic constraints. In earthquake areas, these local traditions were shown to have, in some areas, significant mismatches with the need to build in a way that best resists ground-shaking. In particular, the heavy stone masonry and mud-building traditions of arid areas have led to buildings which are brittle and prone to collapse. The lighter-weight timber-framed buildings of the hot and humid tropics tend to have good earthquake resistance.

This review of residential building patterns in the world's areas of highest seismicity has revealed that, in the rural areas of the poorer countries, in spite of their climatic suitability, timber-framed buildings have been declining in number over recent decades, often to be replaced by more vulnerable masonry buildings. This is partly because of the influence of urban building methods, coupled with a decline in the availability of timber, and in some countries a ban on its use for building, and a corresponding decline in the skills required to construct with it and keep the buildings maintained over their lifetime.

Masonry appears today to be established as the dominant form of construction in the poorer countries. Some country building programmes and regulations are promoting various forms of combined masonry and reinforced concrete construction, in confined masonry or 'restrained brick' buildings. Unfortunately, these are more expensive than unreinforced masonry, and are not yet reaching the poorest rural areas, where unreinforced masonry, in some areas in the form of adobe construction, is still the norm. As a result, a substantial part of the population is left particularly vulnerable.

By contrast, timber-frame construction has become the norm, not in those areas where the climate especially favours it, but in the wealthier countries with an abundance of renewable resources, including timber, and history and public awareness of earthquakes. This is particularly true in Japan, New Zealand and in California, where good quality and safe construction is now ensured by regulations and building supervision.

In dense urban areas, where multistorey construction is required, concrete frame buildings are the most common form of construction. In some well-regulated countries or cities, these are of relatively safe shear-wall construction, but masonry-infilled concrete frame construction is widely practised in urban and suburban areas on a global scale, and poses a great danger to millions of occupants of such buildings.

Efforts have been made in several countries with a masonry or earthen building tradition to develop and teach ways of strengthening traditional buildings. These efforts, which will be discussed in Chapter 7, have tended to be in the context of post-disaster reconstruction programmes and are not yet widespread.

Earthquake safety cannot be considered separately from other concerns affecting the future patterns of house-building. The need to limit energy consumption and greenhouse gas emissions should in future encourage the use of buildings built using methods derived from the vernacular traditions, using low-energy building materials and passive heating and cooling. This issue too will be discussed in later chapters.

References

Alesch, D. and Petak, W. (1986). *The Politics and Economics of Earthquake Hazard Mitigation: Unreinforced Masonry Buildings in Southern California*. University of Colorado.

Arnold, C. (2002). *Wood Frame Single Family House*, Report 65. USA: World Housing Encyclopedia.

Beattie, G. and Thurston, S. (2006). Changes to the seismic design of houses in New Zealand. *Presented at the NZSEE Conference*, New Zealand Society for Earthquake Engineering, New Zealand.

Blondet, M., Torrealva, D., Velasquez, J., and Tarque, N. (2006). Seismic reinforcement of adobe houses using external polymer mesh. *Presented at the First European Conference on Earthquake Engineering and Seismology*, Geneva.

Blondet, M., Villa Garcia, G.M., Brzev, S., and Rubinos, A.W. (2011). *Earthquake-Resistant Construction of Adobe Buildings: A Tutorial*. Earthquake Engineering Research Institute.

Boen, T. (2016). *Learning from Earthquake Damage: Non-Engineered Construction in Indonesia*. Indonesia: Gadjah Mada University Press.

Bothara, J., Ingham, J., Dizhur, D., 2018. Earthquake risk reduction efforts in Nepal, in: Samui, P., Kim, D. and Ghosh, C. *Integrating Disaster Science and Management*. Elsevier.

Buchanan, A., Carradine, D., Beattie, G., and Morris, H. (2011). Performance of houses during the Christchurch earthquake of 22 February 2011. *Bulletin of the New Zealand Society for Earthquake Engineering* 44: 342–357.

Comerio, M. (1998). *Disaster Hits Home: New Policy for Urban Recovery*. Berkeley, CA: University of California Press.

D'Ayala, D. and Speranza, E. (2002). *Single-Family Stone Masonry House*, Report 28. Italy: World Housing Encyclopedia.

Dowrick, D. (1987). *Earthquake Resistant Design for Engineers and Architects*, 2e. Wiley.

Dudley, E. (1993). *The Critical Villager: Beyond Community Participation*. London and New York: Routledge.

Duly, C. (1979). *The Houses of Mankind*. Thames and Hudson.

EEFIT (1990). *The Luzon Philippines Earthquake of 16 July 1990: A Field Report by EEFIT*. London: Institution of Structural Engineers.

EEFIT (2008). *The Wenchuan Earthquake of 12th May 2008: A Preliminary Field Report by EEFIT*. London: Institution of Structural Engineers.

EERI (1994). *Preliminary Report – Northridge, California, Earthquake of January 17, 1994*. California: Earthquake Engineering Research Institute.

EERI (2008). *The Wenchuan, Sichuan Province, China, Earthquake of May 12, 2008*. California: Earthquake Engineering Research Institute.

Gülkan, P., Aschheim, M., and Spence, R. (2002). *Reinforced Concrete Frame Building with Masonry Infills*, Report 64. Turkey: World Housing Encyclopedia.

Hicylmaz, K., Bothara, J., and Stephenson, M. (2011). *Dhajji Dewari*, Report 146. Pakistan and India: World Housing Encyclopedia.

Jain, S.K. (2016). Earthquake safety in India: achievements, challenges and opportunities. *Bulletin of Earthquake Engineering* 14: 1337–1436.

Jaiswal, K. and Wald, D.J. (2014). *Creating a Global Inventory for Earthquake Loss Assessment and Risk Management (Created in 2008, Revised 2014)*. No. Open File Report 2008-1160. Reston: US Geological Survey.

Jaiswal, K.S., Wald, D.J., Earle, P.S. et al. (2011). Earthquake casualty models within the USGS Prompt Assessment of Global Earthquakes for Response (PAGER) system. In: *Human Casualties in Earthquakes* (eds. R. Spence, E. So and C. Scawthorn), 83–94. Springer.

Jofre, D., Abrahamczyk, L., and Schwarz, J. (2011). *Confined and Internally Reinforced Concrete Block Masonry Building*, Report 161. Guatemala: World Housing Encyclopedia.

Koenigsberger, O.H., Ingersoll, T.G., and Mayhew, A. (1973). *Manual of Tropical Housing and Building: Climate Design*. London: Hong Kong Press.

Kottek, M., Grieser, J., Beck, C. et al. (2006). World map of the Köppen–Geiger climate classification updated. *Meteorologische Zeitschrifte* 15: 259–263.

Langenbach, R. (2015). What we learn from vernaclar construction. In: *Nonconventional and Vernacular Construction Materials: Characterisation, Properties and Applications* (eds. K.A. Harries and B. Sharma). Woodhead Publishing.

Lee, W.V. (2008). Mismatches in climatically and seismically dominant built forms. MPhil Dissertation in Architecture. Cambridge University.

Lopez, M., Bommer, J., and Mendez, P. (2004). The seismic performance of bahareque dwellings in El Salvador, paper 2646. *Presented at the 13th World Conference on Earthquake Engineering*, Vancouver.

Maki, N. and Tanaka, S. (2002). *Single-Family Wooden House*, Report 86. Japan: World Housing Encyclopedia.

Mehrain, M. and Naeim, F. (2004). *Adobe House*, Report 104. Iran: World Housing Encyclopedia.

Oliver, P. (1987). *Dwellings: the House Across the World*. UK: Phaidon Press.

Rodriguez, M. and Jarque, F. (2005). *Reinforced Concrete Multistory Buildings*, Report 115. Mexico: World Housing Encyclopedia.

Rubel, F. and Kottek, M. (2010). Observed and projected climate shifts 1901–2100 depicted by world maps of the Koppen–Giger climate classification. *Meteorologische Zeitschrifte* 19: 135–141.

Schilderman, T. (2004). Adapting traditional shelter for disaster mitigation and reconstruction: experiences with community-based approaches. *Building Research and Information* 32: 414–426.

So, E. (2009). The assessment of casualties for earthquake loss estimation. PhD thesis. Cambridge University.

Spence, R. and Cook, D. (1983). *Building Materials in Developing Countries*. Wiley.

Tassios, T.P. and Syrmakezis, K. 2002. Load-bearing stone masonry building, Report 16. Greece: World Housing Encyclopedia.

Tena-Colunga, A., Juarez-Angeles, A., and Salinas-Vallejo, V. (2011). *Report # 160: Combined and Confined Masonry Construction*. World Housing Encyclopedia.

Wan, L. (2013). Study of built environment sustainability of poor rural areas of Southwest China. PhD thesis. Chinese University of Hong Kong.

Yakut, A. and Gülkan, P. (2003). *Tunnel form Building*, Report 101. Turkey: World Housing Encyclopedia.

Yepes-Estrada, C., Silva, V., Valcárcel, J., and Acevedo, A.B. (2017). Modeling the residential building inventory in South America for seismic risk assessment. *Earthquake Spectra* 33: 299–322.

4

What Happens in an Earthquake?

In this chapter, we are going to briefly explain what an earthquake is and examine what happens in an earthquake, to the natural and built environments. Earthquakes that do not cause significant direct building damage or casualties can still be costly due to ground failures such as liquefaction and secondary hazards such as landslides. It is only through understanding the primary, secondary and possible compound risks of earthquakes that we can plan and integrate resilience in communities.

4.1   What is an Earthquake?

Earthquakes are caused by movements in Earth's outermost layer. The Earth's lithosphere is made up of tectonic plates, and these are constantly shifting as they drift over the viscous mantle layer below. When tectonic plates move, it also causes movements at the plate boundaries and an earthquake is the sudden movement of the lithosphere at a fault line. Since friction builds up at the boundaries and the rest of the plate is moving, the energy that would normally cause the plates to slide past one another is accumulated. When the force of the slow-moving blocks finally overcomes the friction of the jagged edges of the fault and releases all the stored energy, an earthquake occurs (USGS 2020a).

The size or magnitude of an earthquake depends on how much the fault has ruptured (the slip) and the area over which the rupture has occurred. When an earthquake occurs, seismic waves are created, which propagate away from the focus or hypocentre, where the earthquake originates. There are seismometers all around the world to help us locate and determine the size of earthquakes. A seismometer is the internal part of a seismograph and is usually in the form of a pendulum or a mass mounted on a spring. A seismograph is securely mounted onto the surface of the earth so that when an earthquake occurs, the entire unit shakes with it except for the mass on the spring, which has inertia and remains in the same place (USGS 2020a). As the seismograph shakes, the recording device on the mass records the relative motion between itself and the rest of the instrument, thus recording the ground motion. These mechanisms are no longer manual but are measured using electronic changes produced by the motion of the ground with respect to the mass. A seismogram is the recording of the ground shaking at the specific location of the instrument

Why Do Buildings Collapse in Earthquakes?: Building for Safety in Seismic Areas,
First Edition. Robin Spence and Emily So.
© 2021 John Wiley & Sons Ltd. Published 2021 by John Wiley & Sons Ltd.

and tells us the size of the event, with time on its horizontal axis and ground displacement on its vertical axis.

There are four main types of earthquake waves: P-waves and S-waves (which are body waves), and Rayleigh waves and Love waves (which are surface waves). Body waves can travel through the Earth's interior, whereas surface waves can only move along the surface of the earth like ripples on water. Travelling only through the crust, surface waves are of a lower frequency than body waves and are easily distinguished on a seismogram as a result. Though they arrive after body waves, it is surface waves that are almost entirely responsible for the damage and destruction associated with earthquakes. Love waves tend to cause the most damage due to their large amplitude. The damage and strength of the surface waves are reduced in deeper earthquakes.

The fastest waves are the P-waves. P-waves also referred to as primary or compressional waves can move through solids, liquids and gases. A P-wave pushes and pulls the rock, it moves through just like sound waves push and pull the air. These are used by seismologists to locate the earthquakes, and they travel at a speed of about 5–8 km/s. S-waves travel more slowly than the P-waves and can only travel through solids, at about 3–5 km/s. The crucial difference in arrival times between P and S waves is sometimes used as a short-term earthquake alert; for example, it is used to automatically stop all Shinkansen trains in Japan ahead of the more damaging S-waves.

Earthquakes are measured in magnitudes and intensities. Intensity measures the level of shaking at a particular location. The Modified Mercalli Intensity (MMI) Scale is widely used and is based on observable earthquake damage, which can be subjective. Therefore, magnitude is the most common measure used. Magnitude provides the size of the earthquake source and is the same number no matter where you are or the level of shaking.

The Richter magnitude scale is a popular but outdated method for measuring magnitude as it only measures the largest amplitude of an event and not the different parts of the earthquake. However, it is commonly used to communicate earthquake information to the public and the familiar understanding of a logarithmic magnitude scale comes from Richter.

A more uniformly applicable extension of the magnitude scale, related to the total energy released by the earthquake was developed in the 1970s, known as moment magnitude, or Mw. To determine the moment magnitude of an earthquake, seismologists measure the strength of the rock along the fault, the area of the fault that has slipped, and the distance the fault has moved. The moment magnitude provides an estimate of earthquake size that is valid over the complete range of magnitudes, a characteristic that was lacking in other magnitude scales, including the Richter scale.

While some earthquakes will have foreshocks (earthquakes happening at approximately the same location as the larger event that follows), all earthquakes will have aftershocks. Sometimes earthquakes happen in swarms with many sporadic low-magnitude tremors, like the earthquakes that preceded the Mw6.3 L'Aquila earthquake in 2009 for four months. The magnitude of an earthquake depends on the fault geometry and the maximum amount of energy that can be stored and released in a single segment (or sometimes multiple segments) at any one time. Depending on the size of the mainshock, aftershocks can continue for weeks, months, and even years after the mainshock.

An earthquake can occur at any depth within the crust of the Earth. Shallow earthquakes of any given magnitude tend to be more damaging than deeper earthquakes since the shock from deeper events must travel further to the surface, losing energy along the way. Although deep earthquakes may be less damaging, their energy can travel much longer distances. The deepest events are associated with subduction zones, when the plates collide, and one subducts beneath another. The largest recorded earthquakes have occurred where oceanic plates collide and dive beneath the less dense continental plates, as shown in the top left image of Figure 4.1. These contact points along the plate boundary are very large and can give rise to very large magnitude earthquakes, for example the Mw9.1 Indian Ocean earthquake and tsunami in 2004, and the Mw9.1 Great Tohoku earthquake in 2011 are two examples of subduction zone earthquakes.

The movements at plate boundaries and faults vary. In addition to subduction zone boundaries, there are three other types as shown in Figure 4.1.

Boundaries between plates are made up from a system of faults. Each type of boundary is associated with one of three basic types of fault, called normal, reverse and strike-slip faults. Normal faults are associated with divergent plate boundaries shown in the bottom left image of Figure 4.1, like the Mid-Atlantic ridge. The Mid-Atlantic Ridge is a submarine ridge located along the floor of the Atlantic Ocean. It is the longest (>65000 km) and the most extensive chain of mountains on earth, but more than 90% of this mountain range remains hidden from view underwater. There are only a few places on earth where it surfaces in the form of a few islands, one of which is Iceland (Figure 4.2).

Reverse faults (or thrust faults) are found at convergent boundaries, shown on the top right image of Figure 4.1. These are associated with mountain ranges, such as the Himalayas which are still forming. Strike-slip faults occur at transform fault boundaries (Figure 4.1, bottom right). Surface rupture occurs when movement on a fault deep within the earth

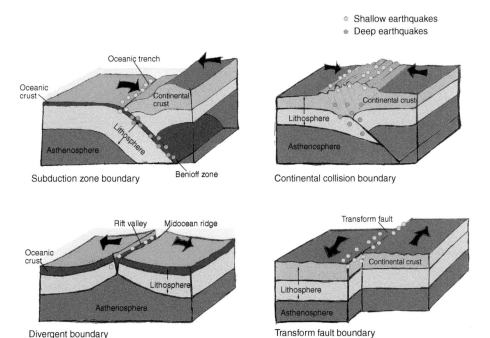

Figure 4.1 Diagrams showing different types of earthquakes.

Figure 4.2 Þingvellir National Park in Iceland, where the Eurasian and North American tectonic plates meet. *Source:* Andrew Schaffer.

breaks through to the surface, though not all earthquakes result in surface ruptures. One of the most spectacular visible ruptures is that of the San Andreas Fault (Figure 4.3), a system of strike-slip faults that makes up the transform boundary of this famous fault in California, USA.

Fault systems create suitable habitats for people. The precise reason why there is a sudden change in topography is the presence of a fault. The topography leads to formation of soils and faulting creates natural harbours. Human settlements in fault zones are common as the break in geology also helps draw up ground water or create underground reservoirs we need to survive. Figure 4.4 has been taken from Jackson (2006). It shows how the water table is elevated nearer to the ground by a thrust fault and how tunnels called 'qanats' have existed since the ancient Persian civilization in Iran to extract this precious resource.

Flat valleys are favourable locations for dwellings as they are protected by mountains against the elements. These are what Jackson had termed a 'fatal attraction'. Today, these towns and villages in valleys along old trade routes have grown into some of the most populous cities in the world, such as Tehran, Quito and Kathmandu. The hazard itself has not changed, but the exposure and therefore risk have significantly increased over time due to urbanisation and increase in population, even with the advances of building technology (see Chapter 8).

Though most earthquakes happen at boundaries between tectonics plates and along faults, there have been some notably large intra-plate earthquakes in the past. The series of seven earthquakes of magnitudes M6.0–7.5 occurring in the period 16 December 1811 through 7 February 1812 in New Madrid, Missouri, USA, and the Mw7.7 Bhuj earthquake in 2001 are examples of intraplate earthquakes which caused significant destruction.

Figure 4.3   The San Andreas fault in California, USA. *Source:* U.S. Geological Survey.

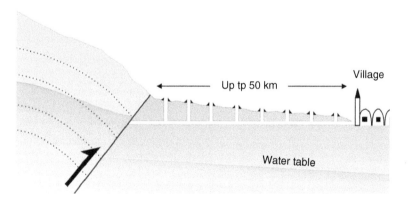

Figure 4.4   An irrigation tunnel (qanat) has been dug through alluvium towards a range-front, where the water table is elevated because of impermeable clay ('gouge') on a thrust fault. *Source:* Jackson 2006. © 2006, The Royal Society.

4.2   Volcanic Earthquakes and Induced Seismicity

These earthquakes are not explicitly covered in this book. Many processes in and around volcanoes can generate earthquakes, which have very different characteristics from the deep and tectonic seismic events that occur along great faults. They are usually much less devastating and may continue at a detectable level for weeks or months prior to eruption.

Induced seismicity from gas extraction and fracking has been a cause for concern in recent years. The extraction of natural gas from regions such as Groningen in The Netherlands and fracking in the United States have been responsible for increased seismicity in the local areas with earthquakes of low magnitudes. The largest earthquake induced by fluid injection that has been documented in scientific literature was a magnitude 5.8 earthquake on 23 September 2016 in central Oklahoma (USGS 2020b).

4.3   How Earthquakes Travel through Different Media

Seismic waves travel faster through hard rocks than through softer rocks and sediments. As the waves pass from deeper and harder rocks to shallow softer rocks, they slow down and get bigger in amplitude as the energy accumulates. The local soil acts as a filter that changes the frequency content and duration of the incoming seismic motion. The different composition and thicknesses of soil layers can either amplify or de-amplify ground motions. Soft soils commonly amplify motions relative to bedrock and the softer the rock or soil under a site, the larger the amplification of the wave.

Local site conditions underlying buildings can cause significant variability in earthquake ground motions experienced by the structures. Figure 4.5 shows two diagrams of relative peak ground accelerations in different US cities. As shown, when only considering bedrock (top), the cities in the West Coast have higher relative peak ground accelerations (PGAs). However, if we consider local site conditions, the bottom diagram shows that soil in Central and Eastern United States increases the seismic hazard, equating potential PGAs in Memphis to those in Seattle and San Francisco.

Site effects in the form of amplification or de-amplification have been observed in many recent seismic events including the Mw8.0 Mexico City earthquake in 1985, the Mw6.7 Loma Prieta earthquake in 1994, the Mw6.9 Kobe earthquake in 1995 and the Mw7.6 Kocaeli earthquake in 1999. The Mw8.0 Mexico City earthquake in 1985 caused severe damage in the capital, which was approximately 400 km away from the epicentre in the Pacific Ocean. The seismic waves were severely amplified inside the lake-bed zone of the city (Humphrey and Anderson 1992). Other densely populated cities with deep soft soil effects around the world include Kathmandu, Los Angeles, Bucharest and Taipei.

4.3.1   Response of Buildings to Earthquakes

What can make earthquakes so devastating to buildings is the way some structures respond to types of ground motion – especially ground accelerations, the predominant frequency of the shaking and its duration. Buildings shake when the frequency of the seismic waves is close to the natural frequency of vibration of the building, an effect known as resonance.

The acceleration and speed of a building's movements will depend on the size of the earthquake, but the period of the building's motion will be the same regardless of the earthquake's size. The natural periods of vibration of a building depend on its mass and its stiffness.

Figure 4.6 shows the ranges of natural periods of structures of different heights. A stocky building would have a natural period of less than one second, a midrise building of 5–15

(a)

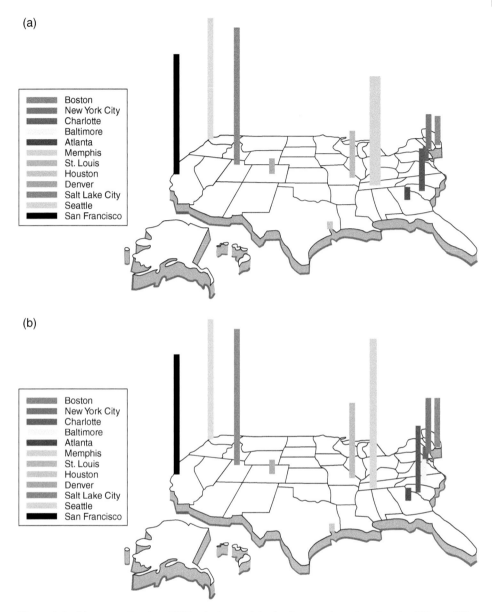

(b)

Figure 4.5 Diagrams showing (TOP) relative peak bedrock ground accelerations in different US cities and (bottom) the resulting surface accelerations after soil amplification effects. Most notable of these are Memphis and New York City. *Source:* FEMA.

storeys would have a period of around a second and tall buildings of >20 storeys, greater than two seconds. If the period of the seismic ground motion matches the natural resonance of a building, it will undergo the largest oscillations possible and suffer the greatest damage. Long period motions from far afield earthquakes would affect tall buildings with natural periods of >2 seconds the most. Although the high-frequency seismic waves would

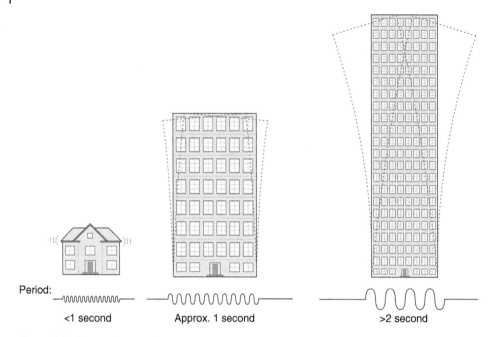

Period:

<1 second   Approx. 1 second   >2 second

Figure 4.6 Figure showing the natural periods of structures of different heights. *Source:* Adapted from IRIS.

lose its energy through long distances, long period seismic waves undergo less energy dissipation and hence can carry energy over longer distances. Typically, these ground motions have very low peak ground amplitudes but long dominant periods.

Earthquake loads are predominantly horizontal, but there is also a vertical component to seismic motions. Observed damage has been attributed to strong vertical motions in buildings and bridges after some important earthquakes in the past including the Mw6.7 Northridge earthquake in 1994, Mw6.9 Kobe earthquake in 1995 and more recently in the Mw7.6 Kashmir earthquake in 2005.

4.3.1.1 Case of Mexico City

An important part of Mexico City, the Central Business District, is built on deep, soft soil that was once the bottom of a lake and reclaimed over several centuries. Instead of cushioning the city from earthquakes, it exaggerates their effects. The seismic waves of earthquakes are amplified by the soil and sediments above, making the surface and the structures built on the surface shake longer and more intensely. Professor James Jackson at the University of Cambridge likens it to jelly – 'It's like being built on jelly on top of something that is wobbling'. There have been two damaging earthquakes in the city in recent years, 1985 and 2017. The Mw7.1 Puebla earthquake in 2017 was called a 'déjà-vu' occurring on the 32nd anniversary of the Mw8.0 Mexico City earthquake in 1985.

More than 39 000 people died and nearly 10 000 buildings collapsed in 1985. This was attributed to a multi-resonance of seismic incident waves, soil deposits, and structures.

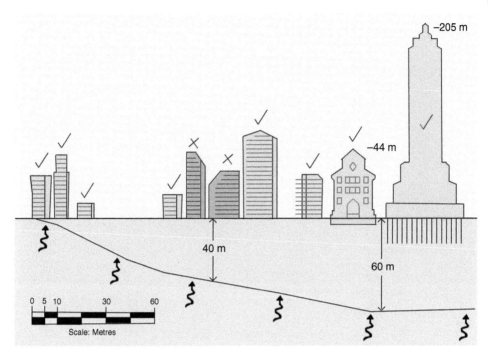

Figure 4.7 This diagram shows the most damaged types of buildings in the 1985 event, denoted by the red and crosses. Damage was related both to building height and also to depth of soft lakebed soils. *Source:* Adapted from Nikolaou et al. 2018.

Most buildings with severe damage were those with 6–15 storeys; the resonant frequency of these buildings was the frequency most amplified by the subsoils in the city.

As shown in Figure 4.7, damage was concentrated in buildings highlighted in red due to strong motions affected by soil resonance at T = 2 seconds, and full excitation was between 1.5 and 2 seconds. The other building types at sites that were not underlain by 40 m of softer soils did not suffer significant damage.

In 2017, the soil amplification was repeated from a different type of earthquake that was 120 km away. The underlying soft soil of the city increased the bedrock seismic motion above the building code acceleration levels and severely affected the buildings with a natural period of 0.8–1.5 seconds in the city (Mayoral et al. 2019). The building codes were changed after the 1985 earthquake, and the code now requires buildings to be built to a peak spectral acceleration of 0.4 g. However, the actual recorded peaks were 0.6 g for the 2017 earthquake, affecting five to eight storey buildings most extensively (Mayoral et al. 2019).

4.3.2 Liquefaction

Liquefaction occurs when saturated soil loses its strength due to strong ground shaking.

> 'If a saturated sand is subjected to ground vibrations, it tends to compact and decrease in volume. If drainage is unable to occur, the tendency to decrease in volume results

Figure 4.8 Clean-up operations of damage caused by liquefaction in Christchurch, New Zealand. *Source:* Brett Phibbs, NZ Herald.

in an increase in pore pressure. If the pore water pressure builds up to the point at which it is equal to the overburden pressure, the effective stress becomes zero, the sand loses its strength completely, and liquefaction occurs' (Seed and Idriss 1982).

Though the number of casualties may be low compared to building collapses and other secondary hazards, the economic and social costs of soil failures are surprisingly high. According to the Association of Bay Area Government, after the Mw6.7 Northridge earthquake in 1994, homes damaged by liquefaction or ground failure were 30 times more likely to require demolition than homes only damaged by ground shaking. After the Mw6.9 Kobe earthquake in 1995, there was significant damage to port facilities due to liquefaction. About 90% of the 167 berths were reported as damaged, and all operations were shut down. An initial estimate at the time suggested that it would take three years to return to full operation, at a cost of US$7bn (EEFIT 1997).

After the Mw6.1 Christchurch earthquake in 2011, liquefaction caused significant economic loss. Although the damage to building structures was not extensive, the liquefaction led to eventual demolition of many buildings as these became inhabitable due to subsidence (Figure 4.8). In addition, the underground pipe network was severely damaged. One study claimed that the lateral and vertical ground movements from liquefaction led to greater damage than the ground motions (Giovinazzi et al. 2015).

4.3.3 Lateral Spreading

Lateral spreading is mostly horizontal deformation of gently sloping ground (<5%) resulting from soil liquefaction. It is one of the most widespread forms of ground damage which does not cause loss of life but is especially damaging (and costly) to roads and lifelines.

4.4   Secondary Hazards

According to a recent study by Daniell et al. (2017), examining historical earthquake losses from 1900, around 100 key earthquakes (or 1% of damaging earthquakes) have caused close to 93% of fatalities globally. In addition, they also found that within these events, secondary effects such as landslide and tsunamis have caused nearly 40% of economic losses and fatalities as compared to shaking effects. The graphs in Figure 4.9 taken from the study show the breakdown of direct and total economic costs of over 9000 events in the CATDAT database and the contribution of these costs from secondary hazards.

Understanding secondary hazards is therefore of utmost importance in our quest to mitigate these losses in the future.

4.4.1   Landslides

Ground shaking is one of the main causes of large landslides. The largest earthquakes are capable of triggering thousands of landslides simultaneously causing considerable damage and loss of life, especially when soil is saturated following heavy rain. Notable recent landslide-generating earthquakes include Mw7.6 Kashmir earthquake in 2005, Mw7.9 Wenchuan earthquake in 2008 and the Mw6.1–7.0 Kumamoto earthquakes in 2016.

The Kumamoto earthquakes, which comprised a main shock and a foreshock, were a series of earthquakes that struck beneath Kumamoto City, Kumamoto Prefecture in the Kyushu Region of Japan, resulting in 69 deaths, nine of which were caused by landslides. The foreshock, a Mw6.1 earthquake struck the region on 14 April 2016 (21:26 local time). This earthquake caused intense shaking in the eastern part of Kumamoto Prefecture, and major earthquake damage was caused in Mashiki Town near the epicentre. Subsequently, on 16 April 2016, at 01:25 local time, a larger Mw7.0 event occurred which caused significantly greater damage in wider areas near the fault, e.g. Mashiki Town, Nishihara Village, and Minami Aso Village (Goda et al. 2016). A massive landslide caused by the main shock collapsed the Aso Bridge as shown in Figure 4.10 and killed a 22-year-old college student

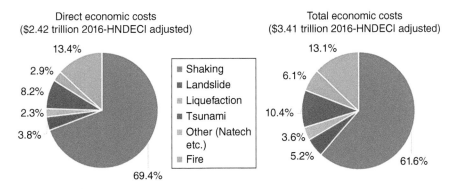

Figure 4.9   Disaggregation of shaking and secondary effects economic costs from 9920 earthquakes from 1900 to 2016 – left: direct economic costs; right: total economic costs. *Source:* Daniell et al. 2017. 2017, Frontiers Media. Licensed under CC BY 4.0.

on his way home from delivering food and water to his friends who were recovering from the foreshock. The Aso Bridge opened in 1971 and was 200 m in length, 70 m above the bottom of the ravine. The collapse of the bridge isolated about 1000 people. The earthquake striking in the middle of the night saved many lives from this one landslide as there would have been many more people and cars on the bridge at other times of the day.

Earthquake-induced landslides can dramatically change the natural landscape as shown from an aerial view of the affected area after the Mw6.6 Hokkaido Eastern Iburi earthquake in 2018, Figure 4.11.

However, unlike the Japanese events shown above, some earthquake induced landslides have led to significant life loss and building damage. The Mw7.9 Wenchuan earthquake in 2008 triggered more than 200 000 individual events and over 800 of them blocked a river (Fan et al. 2018). Landslides, rockfalls and debris flows resulted in about 20 000 deaths (Yin et al. 2009). Entire settlements were buried including a large part of the town of Beichuan, as shown in Figure 4.12.

Apart from direct deaths and damage to the built and natural environment, landslides block road networks and therefore can hamper search and rescue efforts. A remote sensing study (Zhang et al. 2010) revealed that the Wenchuan earthquake induced 170 km^2 of landslides in the southern half of the Longmenshan Fault Zone, most of which was formerly covered by forest or shrub. The local governments organised officials and local people to undertake immediate rescue activities and armed police corps and the People's Liberation Army were also involved in the huge search and rescue efforts, which included clearing of

Figure 4.10 The massive landslide that severely damaged the highway and collapsed the Aso Bridge after the mainshock of the Kumamoto sequence in 2016. At the centre of the image, the road that led to the collapsed bridge on the near side of the valley is visible. *Source:* AP.

Figure 4.11 View of the landslides in the affected area caused by the Mw6.6 Hokkaido Eastern Iburi earthquake in 2018. *Source:* Asahi Shimbun.

Figure 4.12 Landslide in the town of Beichuan after the Mw7.9 Wenchuan earthquake in 2008 in China. *Source:* Author owned image.

roads and airlifting trapped survivors from collapsed structures and 'isolated islands'. The town of Yingxiu close to the epicentre of the earthquake was completely cut off for days and the only access was by foot or helicopter. Helicopters could only rescue 10–12 people at a time and missions were subject to weather conditions. In this one town, over 60% of its

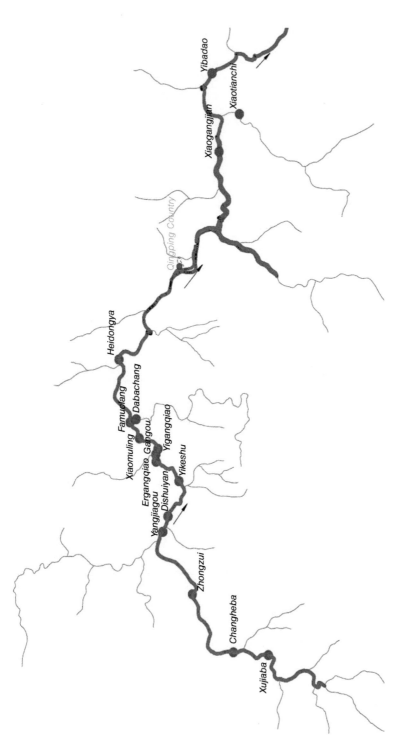

Figure 4.13 Image showing the quake lakes that formed along the Mianyuan River after the 2008 Wenchuan earthquake in China. *Source:* Zhou et al. 2012.

9000 population died. In total, 1.49 million people were rescued from isolated earthquake-affected mountain areas, cut off by the mass landslides (Jiang et al. 2008).

'Quake-lakes' were also a significant secondary hazard of the Mw7.9 Wenchuan earthquake in 2008, where the debris brought down by landslides blocked numerous sections of rivers and hundreds of quake lakes formed. Figure 4.13 shows the quake lakes that formed along the Mianyuan River. These lakes interact with each other, posing great risks of flooding to the villages downstream.

The total volume of landslide dams and river siltation in the upstream areas of the Mianyuan River was almost 50 mil m^3 and the lakes, when fully filled with water were more than 20 mil m^3 in volume. Aftershocks, landslides and water accumulation continued to be a threat for weeks after the main earthquake event in 2008. Some 200 relief workers were reported to have died in mudslides whilst trying to open channels to drain the quake lakes. Breaches to these lakes would have caused serious damage to dwellings and to millions of residents downstream; many villages had to be evacuated.

4.4.1.1 Tsunamis

Tsunamis are giant waves caused by earthquakes or volcanic eruptions under the sea. Offshore earthquakes can trigger tsunamis as they displace large masses of water at depth and energy is transmitted through waves to shallow waters, finally reaching the coastal areas at great speed and height. As the waves approach the coast of a continent, however, friction with the rising sea floor reduces the velocity of the waves. As the velocity lessens, the wavelengths become shortened and the wave amplitudes (heights) increase. A tsunami wave can travel as fast as 800 km/h across the ocean and reach a height of 30 m when it arrives at the coast (FEMA 2020).

The most devastating tsunamis in the past two decades are the Mw9.1 Indian Ocean earthquake and tsunami in 2004 and the Mw9.1 Great Tohoku earthquake in 2011 along the east coast of Japan. Since there is a time lag between the earthquake and tsunami waves reaching land, warnings and evacuations are possible. The dedication and alternative tsunami education programme of Professor Toshitaka Katada in Kamaishi prior to the Mw9.1 Great Tohoku earthquake in 2011 (Box 4.1) meant that virtually all the town's school-age children survived the tsunami because they evacuated in time.

Tsunamis happen most frequently in the Pacific Ocean because of the many large earthquakes associated with subduction zones along the margins of the Pacific Ocean Basin, which is called the 'Ring of Fire'. Other areas of high risk as identified in the map in Figure 4.14 include Portugal, the Mediterranean and India.

In many coastal areas around the Pacific Ocean: Chile and Peru, West Coast USA, Japan, and New Zealand, tsunami drills are commonly practised. Emergency plans and evacuation can reduce the loss of life significantly but not the immense impact on the built and natural environment. Both the 2004 and 2011 events caused immense losses to buildings and settlements through wave damage (see examples in Chapter 2). Tsunami waves cause damage to the landscape, bringing in debris from the sea and along its path on land. Furthermore, the sea water ingress causes salination of agricultural soil and flooding creates problems with transportation, power, communications and drinking water.

Box 4.1   Profile: Toshitaka Katada

Toshitaka Katada: The Miracle of Kamaishi

Toshitaka Katada is professor of civil engineering at Tokyo University in Japan. In the tsunami which followed the Mw9.1 Great Tohoku earthquake in 2011, over 18 000 lives were lost, mostly by drowning, including over 1000 in the small coastal town of Kamaishi in Iwate Prefecture. Yet, thanks to the tsunami disaster prevention education programmes that Katada had planned and carried out in the schools of Kamaishi, virtually all the town's school-age children survived the tsunami because they evacuated in time. The evacuation has been called 'The Miracle of Kamaishi'. How did this come about?

As a flood prevention specialist, Katada came to focus on tsunami disaster prevention as a result of the devastating consequences of the Mw9.1 Indian Ocean earthquake and tsunami in 2004. Touring the areas devastated by the tsunami, he was particularly struck by the story of Mai Khao Beach in Thailand. On that beach 10-year old Tilly Smith, from UK, was holidaying with her family on the day the tsunami struck. Seeing the water recede far below normal low tide level, she realised that this was exactly the tsunami warning sign she had been taught about in a geography lesson at her school only three weeks earlier. Not waiting for adult confirmation, Tilly had shouted at all those around her that a tsunami was coming and that they should to move to higher ground. Almost 100 people followed her lead, and as a result it was one of the few stretches of beach where nobody drowned (Muir-Wood 2016).

Katada was concerned that even though Japan's coastal regions had been warned of a possible major earthquake, the alert level among the people was low. Also, he discovered that children would not evacuate unless an official, or their parents, told them to. As he put it

'Children look at the grown-ups and do what they do. I felt that if those children lost their lives to a tsunami, the fault would lie with not just the parents but adults and society. I knew I had to do something so the children can save their own lives' https://mnj.gov-online.go.jp/kamaishi.html.

He chose to focus on the front-line town of Kamaishi. After some initial resistance, Katada managed to persuade the teachers in all 14 schools in Kamaishi to devote 10 hours a year to tsunami education. Working with the teachers he produced a manual, which emphasised three principles:

1) Don't believe in preconceived ideas: hazard maps showing areas at risk can be wrong.
2) Do everything you can and support others.
3) Take the lead in the evacuation: don't wait for adults to direct you.

Katada thus put the emphasis not just on informing children about tsunamis but developing the right attitude to respond when an event is threatened.

Classroom training session

The test for this education programme came on the afternoon of 11 March 2011, when the gigantic tsunami struck the Kamaishi coast, overwhelming the inadequate coastal protection walls and inundating the town. At that time, children were in school and were able to put their training into practice. Virtually, all schoolchildren evacuated to higher ground, and although the tally of dead or missing for Kamaishi was 1200, every single student who was in school that day survived, in contrast to the experience of neighbouring towns where many children died (Muir-Wood 2016).

Evacuation drills

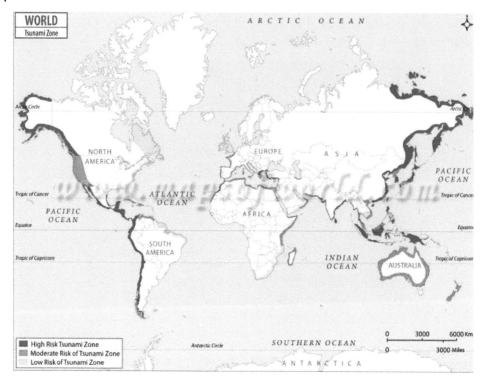

Figure 4.14 Map showing the areas at risk from tsunamis around the world. *Source:* Maps of the world.

4.4.2 Fire-Following Earthquake

For at least four days, 50–60 fires burned and devastated the city of San Francisco after the M7.8 San Francisco earthquake in 1906. The damage resulting from the fires that followed the earthquake in 1906 significantly exceeded the losses from ground shaking (Krezel 2017). Fire-following earthquake refers to a series of conflagration events initiated by a large earthquake. Some notable fires following earthquakes in the last century took place after the Mw8.1 Great Kanto earthquake in 1923, the Mw6.6 San Fernando earthquake in 1971, the Mw8.0 Mexico City earthquake in 1985 and the Mw6.9 Kobe earthquake in 1995 (Figure 4.15).

The most prone areas are regions of high seismicity with a large urban area predominantly comprising of densely spaced timber buildings, such as Japan, New Zealand, Western USA and parts of Southeast Asia, like the Philippines.

The factors affecting the severity of this secondary hazard include the following:

- number of ignitions (ignition rate),
- number of large fires,
- wind speed/direction profile by hour/day during the fire,
- fire load (how much combustible material is available).

Much has been done in earthquake prone areas to mitigate fire-following earthquakes, e.g. automatic gas shut off valves, as mandated in Japan after Mw6.9 Kobe earthquake in 1995.

Figure 4.15 Fire following the Mw6.9 Kobe earthquake in 1995 in Japan. *Source:* Chiara 2018.

All urban gas meters in Japan have seismic shutoff (Scawthorn 2008) and as a result fire-following earthquake has been in recent decades a smaller hazard than it was. However, the chance of fires getting out of control is still significant. Extrapolating from the 110 fires that occurred after the Northridge, San Fernando and Loma Prieta earthquakes in California, the ShakeOut Scenario (Chapter 7) assumes a Mw7.8 earthquake on the southernmost 300 km of the San Andreas fault will yield 1600 fires. With the sheer number of fires and the likely damage to water supplies and road networks, the chances of fire rescue will be small, consequently residents have been urged to be prepared with ways of extinguishing fires before they become a conflagration, in a part of the world where almost all individual residential homes are made of timber (Scawthorn 2008).

4.5 Compound Threats

The Fukushima Nuclear Power Plant failure in 2011 and the recent Covid-19 pandemic of 2020 have highlighted the need to examine compound risks. The Mw9.1 Great Tohoku earthquake in 2011 triggered a 14 m high tsunami at the Fukushima Daiichi Nuclear Power Plant disabling all AC power to Units 1, 2 and 3 of the Power Plant, and carried off fuel tanks for emergency diesel generators. Despite heroic efforts by the local team, the cooling systems failed, and hydrogen explosions damaged the facilities, releasing a large amount of radioactive material into the environment. The failure of the Daiichi plants not only affected the local areas, where all residents within a zone had to be evacuated, it also sparked a global review of the use and safety of nuclear energy production. The first evacuation, of those within a 2 km radius of the plant, was ordered on the evening of March 11, just hours after the tsunami. The following morning the exclusion zone was expanded to 10 km, but with high radiation levels recorded at the site boundary after the first explosion that day, it was further extended to 20 km around the plant. In all 63 000 people were displaced from the surrounding areas. The consequences of this great earthquake and tsunami for the affected population however go far beyond evacuations. The evacuation, not the nuclear accident itself, was the most devastating part of the disaster say some experts. 'With hindsight, we can say the evacuation was a mistake', says Philip Thomas, a professor of risk management at the University of Bristol. There have been reports of a terrible toll in depression, joblessness and alcoholism among the 63 000 people who were displaced beyond the prefecture; of those, only 29 000 have since returned. In addition to the 18 000 people who lost their lives due to the earthquake and tsunami (see Chapter 2), there have been 2202 disaster-related deaths in Fukushima, according to the government's Reconstruction Agency, from evacuation stress, interruption to medical care and suicide. Of these deaths, 1984 were people over the age of 65. This was a compound disaster.

In Croatia, the Mw5.3 Zagreb earthquake in 2020 occurred at 06:24 local time on Sunday 22, March, with an epicentre 7 km north-east of the city centre. This was quickly followed by a Mw5 earthquake on the same day at 07:01, with an epicentre very close to that of the main shock. The maximum felt intensity from the main shock was reported as VII–VIII on the MCS Macroseismic Intensity Scale, indicating strong shaking. There was one fatality and 26 injuries reported due to the earthquake and another two people died from post-earthquake clear up operations. Although this is not an earthquake of global significance in terms of its reported casualties or damage, what makes this a noteworthy event is its timing. The Croatian capital had been in a partial lockdown in response to the coronavirus pandemic at the time of the earthquake, and there were reports of confusion and panic among residents after the earthquake (Sigmund et al. 2020). In a press conference, the prime minister Andrej Plenković said 'We are urging all citizens to be extra careful. We are recommending staying outside for now. No need for panic... and maintain the distance recommended because of the coronavirus epidemic...We have two crisis situations which contradict each other.' Based on Covid-19 alone, the message was to stay at home. (Walker 2020). On the following day (23 March 2020), Croatia's Covid-19 crisis management team banned all travelling inside the country after many residents of Zagreb fled the city on Sunday morning following the earthquake. The health minister Mr. Vili Beroš, exhorted citizens to still comply with social distancing measures and stated that 'earthquakes are dangerous, but coronavirus is even more so'. As part of this message, Croatians were urged to avoid public squares and parks (Walker 2020). Croatia had 254 confirmed cases of Covid-19 at the time of the earthquake (Sigmund et al. 2020).

The response to this event both by the national civil protection in Croatia and from Europe, as well as the behaviour of the general public have been affected by the pandemic. Dealing with multiple crises with opposing behavioural traits, e.g. keeping social distancing and evacuating affected households to co-living in temporary shelters, prioritising resources and delivering clear public health messages have been some of the challenges the emergency managers and authorities in Croatia faced. The military, emergency services and civilians who helped with the clear up and damage assessments after the earthquake during a lockdown carried out their duties in full acknowledgement of the risks (Z. Koren, 2020, personal communication). Some efforts including the retrieval and analyses of seismological data and damage assessments were hindered and had to be carried out offline or by drones (So et al. 2020).

In an unexpected way, the pandemic may have helped reduce the impact of this earthquake, as the main modes of failure were falling chimneys and ornament damage at roof level, and out of plane gable failures of masonry structures. Due to the time of the day and the pandemic, there were very few people on the streets and no one in churches for morning masses. Twelve hospitals were damaged and partly evacuated in this moderate event. If the earthquake had been a repeat of the M6.3 Zagreb earthquake in 1880 as many seismologists warned, the number of casualties, damage to properties and number of people needing rehousing would have made the compound effect of Covid-19 much worse. The Croatian Chamber of Civil Engineers (HKIG), whose members volunteered to carry out damage assessments immediately after the earthquake was reported to say that 'This earthquake showed us that in the event that Zagreb is struck by an earthquake of magnitude over 6.0, which is unfortunately possible for Zagreb and its surrounding area, the consequences would be catastrophic'.

Planning to mitigate the effects of future catastrophes must always consider the possibility of compound events in which the primary earthquake event triggers or exacerbates other hazards and underlying economic and social conditions. The devastating Mw7.0 Haiti earthquake in 2010 with an estimated death toll of 230 000 is a good example where non-existent or badly enforced building codes due to its unstable economic and political environment turned many buildings into 'weapons of mass destruction' (Bilham 2010). By contrast, the Mw8.8 Maule earthquake in the same year killed just over 700 people in Chile. Tertiary effects such as epidemics, susceptibility of population and climate, socio-psychological and economic status, and environmental susceptibility of a region all play important roles in shaping the severity of consequences from earthquakes.

References

Bilham, R. (2010). Lessons from the Haiti earthquake. *Nature* 463: 878–879. https://doi. org/10.1038/463878a.

Chiara, L. (2018). Fire following earthquake. PhD Dissertation. Department of Architecture and Design Architecture for Sustainable Project, Politecnico di Torino.

Daniell, J.E., Schaefer, A.M., and Wenzel, F. (2017). Losses associated with secondary effects in earthquakes. *Frontiers in Built Environment* 3 https://doi.org/10.3389/fbuil.2017.00030.

EEFIT (1997). *The Hyogo-ken Nanbu (Kobe) Earthqake of 17 January 1995: A Field Report.* London: Institution of Structural Engineers.

Fan, X., Juang, C.H., Wasowski, J. et al. (2018). What we have learned from the 2008 Wenchuan Earthquake and its aftermath: a decade of research and challenges. *Engineering Geology* 241: 25–32. https://doi.org/10.1016/j.enggeo.2018.05.004.

FEMA (2020). FEMA factsheet on tsunamis. http://www.cert-la.com/downloads/education/ spanish/FEMA_FS_tsunami_508-8-15-13.pdf.

Giovinazzi, S., Black, J.R., Milke, M., Esposito, S., Brooks, K.A., Craigie, E.K., Liu, M., 2015. Identifying seismic vulnerability factors for wastewater pipelines after the Canterbury (NZ) earthquake sequence 2010–2011. *Pipelines 2015: Recent Advances in Underground Pipeline Engineering and Construction* 2015, 304–315. doi:https://doi.org/10.1061/9780784479360.029.

Goda, K., Campbell, G., Hulme, L. et al. (2016). The 2016 Kumamoto earthquakes: cascading geological hazards and compounding risks. *Frontiers in Built Environment* 2: 19. https://doi.org/10.3389/fbuil.2016.00019.

Humphrey, J.R. and Anderson, J.G. (1992). Shear-wave attenuation and site response in Guerrero, Mexico. *Bulletin of the Seismological Society of America* 82 (4): 16221645.

Jackson, J. (2006). Fatal attraction: living with earthquakes, the growth of villages into megacities, and earthquake vulnerability in the modern world. *Philosophical Transactions of the Royal Society A: Mathematical, Physical and Engineering Sciences* 364: 1911–1925. https://doi.org/10.1098/rsta.2006.1805.

Jiang, L., Wang, J., and Liu, L. (2008). Providing emergency response to Sichuan earthquake. In: *Technical Assistance Consultant's Report 198*. Asian Development Bank.

Krezel, J. (2017). Modeling fire following earthquake at high resolution. WWW Document. AIRWorldwide. https://www.air-worldwide.com/blog/posts/2017/1/modeling-fire-following-earthquake-at-high-resolution/ (accessed 9 November 2020).

Mayoral, J.M., Asimaki, D., Tepalcapa, S. et al. (2019). Site effects in Mexico City basin: past and present. *Soil Dynamics and Earthquake Engineering* 121: 369–382. https://doi.org/10.1016/j.soildyn.2019.02.028.

Muir-Wood, R. (2016). *The Cure for Catastrophe: How We Can Stop Manufacturing Natural Disaster*. 368 pp. Basic Books.

Nikolaou, S., Gazetas, G., Garini, E. et al. (2018). STRUCTURE magazine|geoseismic design challenges in Mexico City. *Structure Structural Performance*, p. 5.

Scawthorn, C. (2008). *The ShakeOut Scenario Supplemental Study: Fire Following Earthquake*. SPA Risk LLC.

Seed, H. and Idriss, I. (1982). *Ground Motions and Soil Liquefaction During Earthquakes*. Berkeley, CA: Earthquake Engineering Research Institute.

Sigmund, Z., Uroš, M., and Atalić, J. (2020). The earthquake in Zagreb amid the COVID-19 pandemic: OPINION. WWW Document. https://www.undrr.org/news/earthquake-zagreb-amid-covid-19-pandemic-opinion (accessed 9 November 2020).

So, E., Babić, A., Majetić, H. et al. (2020). The Zagreb Earthquake of 22 March 2020: Report for EEFIT. Retrieved from https://www.istructe.org/resources/report/remote-study-zagreb-earthquake-22-march-2020/.

USGS (2020a). Earthquake hazards education. WWW Document. https://www.usgs.gov/natural-hazards/earthquake-hazards/education (accessed 9 November 2020).

USGS (2020b). How large are the earthquakes induced by fluid injection? WWW Document. https://www.usgs.gov/faqs/how-large-are-earthquakes-induced-fluid-injection?qt-news_science_products=0#qt-news_science_products (accessed 9 November 2020).

Walker, S. (2020). *Zagreb Hit by Earthquake While in Coronavirus Lockdown*. The Guardian.

Yin, Y., Wang, F., and Sun, P. (2009). Landslide hazards triggered by the 2008 Wenchuan earthquake, Sichuan, China. *Landslides* 6: 139–152. https://doi.org/10.1007/s10346-009-0148-5.

Zhang, W., Lin, J., Peng, J., and Lu, Q. (2010). Estimating Wenchuan earthquake induced landslides based on remote sensing. *International Journal of Remote Sensing* 31: 3495–3508. https://doi.org/10.1080/01431161003727630.

Zhou, H., Zhang, L., and Yang, X. (2012). Factors influencing breach risk of Quake Lake Group. *Procedia Environmental Sciences, International Conference of Environmental Science and Engineering* 12: 815–822. https://doi.org/10.1016/j.proenv.2012.01.353.

5

How Do Different Forms of Construction Behave in Earthquakes?

5.1 Introduction: Range and Classification of Building Construction Types

In Chapter 3, we looked at the range of building types that are found across the world in the earthquake-prone regions, with the aim of identifying the principal ways of building in each region, and the way in which building form is influenced by the geography, climate, living patterns and economy of each region. In this chapter, we are going to examine in more detail how each of the principal building types responds to earthquakes, what level of earthquake resistance we can expect to find, how buildings of different types are damaged in earthquakes and how and why they may collapse. We will consider how new buildings of each type can be better built to improve their earthquake-resistance, and we will look at strategies which can be adopted for retrofitting existing buildings. The main focus is on the use of each building type for housing, but the principles and methods described can be applied to buildings for any use.

A classification of building construction types is necessary to organise this investigation. There are many building classification systems in use. The purpose here is to classify buildings according to their likely way of responding to an earthquake. The classification system needs to be relatively simple, with not too many different classes, and replicable by different observers; it needs to be internationally applicable; and it should include the vast majority of buildings in use in each region (Brzev et al. 2013).

Here we will use a classification based on that used in the European Macroseismic Scale (Grünthal et al. 1998), the original purpose of which was to enable the measurement of earthquake ground shaking intensity in an earthquake by observing and recording the level of damage to different types of buildings. In an earlier project, we adapted this system for application for a proposed International Macroseismic Scale (Foulser-Piggott and Spence 2013) so that it was more inclusive of building types found outside Europe. Our classification identifies four main groups of construction types, according to the principal structural material used to carry the loads – masonry, reinforced concrete, steel and timber. Within these groups nine sub-groups are identified, according to the ways in which the load-bearing structure is organised. These nine sub-groups are shown in Table 5.1, and the principal characteristics and variants of each sub-group are identified. Comparable data are

Why Do Buildings Collapse in Earthquakes?: Building for Safety in Seismic Areas,
First Edition. Robin Spence and Emily So.
© 2021 John Wiley & Sons Ltd. Published 2021 by John Wiley & Sons Ltd.

Table 5.1 Principal classes of buildings and brief descriptions.

Structural material	Group	Sub-group	Brief description
Masonry	Vernacular unreinforced masonry	Adobe/earthen; rubble-stone	Load-bearing walls of weak masonry, either earthen (adobe, earth bricks or rammed earth), or rubble-stone in lime or mud mortar; roof of timber poles or joists, covered with earth or metal roof sheet; generally single storey.
	Unreinforced load-bearing	Cut stone or brick masonry; concrete block masonry; unreinforced masonry with RC floors	Load-bearing walls of unit masonry, brick, concrete block or stone, laid in courses with mortar of cement or lime; floors of timber joists supporting timber boards, vaults, or reinforced concrete (RC) slab; roof generally pitched and covered with tiles or metal roof sheet, occasionally RC slabs; generally up to three storeys, sometimes up to six storeys.
	Structural masonry	Reinforced or confined masonry	Load-bearing walls of reinforced or confined masonry; floors either of timber joists supporting timber boards or RC slabs; roof generally pitched and covered with tiles or metal roof sheet, occasionally RC slabs; up to four storeys.
Reinforced concrete	Reinforced concrete	Reinforced concrete frame	Loads carried by reinforced concrete moment-resisting frame consisting of beams and columns; frame is either of cast-in-place or precast concrete; infill walls of masonry or other materials; floors and roof generally of reinforced concrete.
	Reinforced concrete	Reinforced concrete shear wall	Loads carried by reinforced concrete bearing wall, or by an infilled reinforced concrete frame with additional regularly spaced reinforced concrete walls; frame is either of cast-in-place or precast concrete; floors and roof generally of reinforced concrete.
Steel	Steel frame	Steel frame	Loads carried by a steel frame, either moment-resisting or braced, with infill walls of a variety of materials; floors and roof of metal deck on steel beams, sometimes reinforced concrete. Single up to multi-storey.
	Steel frame	Light steel frame	Loads carried by lightweight steel frame, widespan, usually prefabricated. Used mainly for retail or industrial buildings; generally single storey.
Timber	Timber frame structures	Timber stud-wall	Loads carried by a timber frame using stud walls with timber cladding or brick veneer; floors and roof of timber joist construction, roofs normally pitched with covering of tiles or metal sheets; generally up to four storeys.
	Timber frame structures	Post and beam	Loads carried by widely spaced post and beam construction with masonry infill or lightweight cladding; mainly low-rise.

not available for all of these sub-groups, but in the following sections, we describe the characteristics, the earthquake behaviour and the typical modes of failure for seven of the nine sub-groups.

5.2 Masonry Construction

5.2.1 Adobe Construction

Construction using earth in the form of adobe blocks is very common in many countries, particularly in South and Central America and South Asia. Although the proportion is now declining somewhat it was estimated in 2009 that around 30–50% of the world's population lived or worked in earthen buildings (Rael 2009). Adobe is the most common earthen building technology. Adobe buildings are suitable for small rural communities because the materials for it are widely available, and because the construction process is simple. Adobe buildings can be and generally are, built by their owners.

Adobe blocks are made from suitably selected earth, which is mixed with water and additives such as straw, cast into block-sized moulds, and air-dried. Walls of adobe buildings are made by laying the blocks in a mud mortar (made from the same earth). Walls are from 250 mm thick upwards, depending on the climate and the need for insulation, and blocks are laid with staggered joints, usually in a single layer or 'wythe'. These buildings usually have flat roofs of compacted earth on timber joists or pitched roofs on timber joists with clay tiles or metal sheet roof covering or sometimes domed roofs (Figure 5.1). Buildings are usually single storey, though they may in some places reach three storeys, and the plan of the building is normally divided by cross-walls so that individual wall elements are of storey height and relatively short.

In spite of the fact that adobe construction is used in many highly seismic areas, its earthquake performance is poor, and many deaths have been caused by the collapse of adobe buildings, as in the Mw7.9 Ancash, Peru earthquake in 1970, the Mw7.7 earthquake in El Salvador in 2001, the Mw6.6 Bam, Iran earthquake in 2003 and the Mw8 Pisco, Peru earthquake in 2007. Typical damage patterns in these buildings have been in the form of cracking

Figure 5.1   Traditional adobe construction in Iran. *Source:* EERI. Reproduced with permission.

and separation of walls at the corners, diagonal cracking of walls, and out-of-plane failure of wall panels. Collapse of walls poorly tied to the roofs which they support frequently leads to complete building collapse (Figure 2.6).

The principal recommendations for unreinforced adobe construction are to

- Choose appropriate soils
- Ensure good wall construction
- Create a robust layout of walls and connections
- Ensure adequate wall-to-roof connections.

Suitable soils will contain a mixture of fine clay material and coarser sandy material. Clay is essential as the binder, but if there is too much clay, shrinkage of the soil will cause cracking and loss of strength. A set of simple tests (Blondet et al. 2011) can be used to find a suitable soil, and sand can be added to soils with too much clay. Adobe walls should be built on a foundation of concrete, brick or stone masonry to limit water penetration from the ground. Vertical and horizontal mortar joints should be completely filled. And adobe walls should be covered with a mud plaster for protection, regularly maintained. Robust layout requires building only single-storey buildings, using a wall layout with intersecting walls at regular intervals (or buttressing) (Coburn et al. 1995), and keeping openings small. A lightweight well-insulated roof rather than a heavy roof of compacted earth or stone should be used. More detailed dimensional recommendations are given in the World Housing Encyclopaedia Tutorial (Blondet et al. 2011).

Even well-built adobe buildings will be vulnerable to intense ground shaking, but research at the Catholic University of Peru (PUCP) and elsewhere (Blondet et al. 2011) has led to a set of recommendations for building unreinforced adobe buildings so as to limit their earthquake vulnerability, and for low-cost techniques for seismic reinforcement both of new and existing adobe buildings.

These techniques include the use of timber ring-beams to tie the tops of walls together and provide improved connection with roof structures, internal reinforcement in adobe walls (both horizontal and vertical) using cane, and external reinforcement with a mesh of cane and rope, or with bamboo or a polymer mesh. Manuals and guide-books for using these techniques have been developed (Blondet et al. 2011), and they have been used in a number of pilot projects. In the Mw8.4 Arequipa, Peru earthquake in 2001, adobe houses built using polymer mesh withstood the earthquake without damage, in areas where similar houses without reinforcement collapsed or were seriously damaged (Blondet et al. 2011; Papanikilaiou and Taucer 2004). External reinforcement techniques can also be used to strengthen existing buildings.

5.2.2 Stone Masonry Construction

Stone construction is often the predominant mode of construction for rural dwellings in parts of the world where suitable stone is available, and where harsh climates demand well-insulated homes which also provide protection from snow, a hot climate or heavy rainfall. Stone is widely used in mountainous regions of South Asia and China, in parts of Iran and Turkey, and in mountainous regions of South America. Stone was also used in earlier centuries for urban multi-storey dwellings as well as rural construction in many parts of Europe such as Italy, Portugal, Croatia, Slovenia and Romania, and many such buildings still exist. Indeed stone masonry construction is still practised today in some parts of Europe.

The form of stone masonry most widely used for rural dwellings has thick walls (upwards of 0.5 m) of undressed stone laid in a mud mortar. Such random rubble masonry buildings are usually single storey, with walls constructed in separate external and internal wythes of stones laid in mortar, with the core filled with loose rubble material. In hot climates, these buildings commonly have a flat roof made from timber poles or joists on which are laid boards or brushwood and a thick surface layer of compacted earth. In some areas, the roof is supported directly on the masonry walls; in others a substructure of timber posts and beams supports the roof independently of the walls. In arid areas, pitched roofs with metal sheet or stone tiles are common (Figure 5.2). Random rubble masonry has had a very poor performance in many earthquakes, such as the Mw7.8 Gorkha, Nepal earthquake in 2015, the Mw7.6 Kashmir, Pakistan earthquake in 2005, the Mw7.7 Bhuj earthquake in 2001 and the Mw6.2 Latur, India earthquake in 1993, and the vast majority of more than 100 000 deaths in those four events has been attributed to collapse of such buildings.

The principal defect of rubble masonry construction when subjected to an earthquake is the lack of structural integrity. This is characterised by a lack of connection between inner and outer wythes of the walls, as well as the same failings as described for adobe buildings, poor connections between walls at their intersections and poor connections between the walls and the roof they support. Vertical cracking at wall junctions and corners is very common, and at higher earthquake intensities, sections of walls can completely disintegrate or overturn (Figure 5.3). The failure of walls leads, as in adobe construction, to the collapse of roofs, the weight of which can result in most occupants being killed or severely injured.

Figure 5.2 Stone masonry construction with roofs of stone tiles, India, Chamoli District.
Source: Rajendra and Rupal Desai. Reproduced with permission.

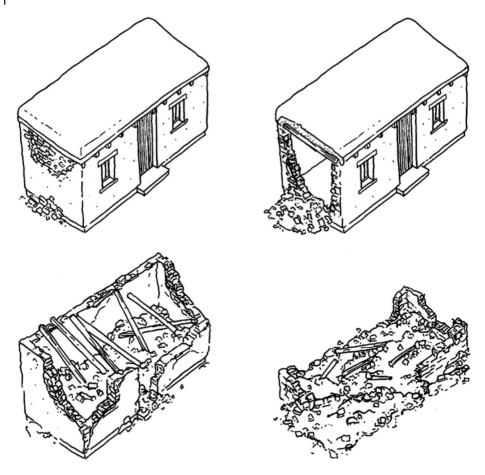

Figure 5.3 Progressive damage states for random rubble stone masonry building. Top left: skin splitting, top right: end wall separation, bottom left: roof collapse, bottom right disintegration. *Source:* Drawing after Andrew Coburn.

The principal ways in which stone masonry buildings can be improved for earthquake resistance are the following:

- to provide integrity within the wall construction and
- to provide structural integrity of the whole building by creating a box action in which walls, floors and roof act together to resist seismic forces.

Integrity within the wall can be achieved by good quality bonding and good mortar throughout the wall, by limiting the height and length of each wall element, by limiting the size and positioning of openings and, crucially, by ensuring that the inner and outer wythes are tied together using through-stones placed at regular intervals. Details and dimensional recommendations are given in the World Housing Encyclopaedia's Stone Masonry Tutorial (Bothara and Brzev 2011). Recommendations specific for particular countries are also given in their national building standards (Bureau of Indian Standards 2003; Government of Nepal 1994a). See also Box 5.1.

Box 5.1 Profile: Laurie Baker

Laurie Baker – sustainable architecture for rural India

Laurie Baker on site in Kerala, 1975. *Source:* Author owned image.

From the late 1940s until his death in 2007, Laurie Baker developed, and applied to the many hundreds of buildings he designed, a unique approach to sustainable architecture. His approach was derived both from an intimate understanding of the local climate, available building materials and craft skills, and from detailed attention to the specific needs of his – often relatively poor or cash-strapped – clients. He thus created an elegant, simple and essentially Indian architecture for the present day, which was in stark contrast to India's western-influenced, resource-intensive architecture of the time.

A lifelong Quaker, Gandhian principles infused his work as they did his life. 'I now think Gandhi was right', he wrote in 1975 'when he said that all building materials should be found within 5 miles of the site'. And 'low-cost techniques should not be considered only for the poor – our aim should be to design only the simplest buildings for all'.

To build cheaply, he ruthlessly pruned all non-local materials. In Kerala, where he lived from the late 1960s, cement plasters were eliminated, and flat reinforced concrete roofs, window glass and bars were replaced by inventive uses of local bricks, clay tiles, timber and lime. Window openings were replaced by patterns of small openings (jali) in the brickwork, providing adequate light, ventilation and security.

This was not achieved without controversy. Personally warm, generous and unassuming, Laurie Baker was outspoken as a professional. When builders refused to adopt his techniques, he recruited and trained his own team of masons. Then, inevitably, his work sparked professional opposition. The engineers of the Public Works Department pilloried his work as a kind of 'loin-cloth' architecture which would not survive long. He, in return, in articles written for the national newspapers, was a savage critic of the derivative 'mock-modern' style of architecture adopted by so many architects in India, for its wastefulness of resources and unsuitability for the climate.

(Continued)

Box 5.1 (Continued)

Baker's confidence to build and write in this manner was developed through the experience of several decades of work as a designer and builder of rural hospitals. Trained as an architect in the United Kingdom, Baker served with the Friends Ambulance Unit in rural China during the war involved in the treatment of leprosy. In 1944, on his way back to England, he had the chance encounter with Gandhi which was to shape his subsequent life and work. Gandhi encouraged him to return to work in rural India, and he went as an architect and anaesthetist for a medical mission in a remote part of the Himalayas. He later wrote 'wherever I went, I saw the indigenous style of architecture, the result of thousands of years of research on how to use only locally available materials. … this was an incredible achievement'.

Most of the architecture for which Baker is properly admired was built in the southern State of Kerala (not an earthquake area) from the 1960s onwards, but as his reputation grew, he was invited to act as advisor to many government committees concerned with housing. Thus, he returned to the Himalayas as advisor to the Department of Science and Technology to study the damage caused by the Mw6.8 Garhwal earthquake in 1991 and recommended what should be done. We have a copy of his hand-written and personally illustrated report on this mission (illustrations below). He demonstrates that most of the damage was caused by poor quality masonry in the stone walls – the traditionally available building material – and could have been avoided by better selection of stone and better bonding of the masonry. Salvaging the better blocks and timber roof beams from the damaged buildings and using local quarries would enable rapid reconstruction and provide safety from all but the largest earthquakes. He wrote 'over a period of nearly two decades the writer experienced several tremors in the area. The whole of Garhwal is used to these tremors and on the whole the traditional house design has withstood this inconvenience'. Rather than a reconstruction programme using imported technology, he recommends training new masons, showing carpenters how to use wood more economically, and applying these techniques in demonstration houses. We do not know how far these recommendations were followed, but they exemplify an approach which has wide application.

Illustrations from Laurie Baker 1991 Report 'a study of the Conditions of the earthquake-hit villages in Garhwal' Left: typical damage, Right: Use of salvaged stones for rebuilding. *Source:* Laurie Baker 1991 Report "a study of the Conditions of the earthquake-hit villages in Garhwal".

The principal means to create seismic integrity for the building as a whole is to provide seismic bands (ring-beams) at lintel, floor and roof level. Ring-beams create continuity between walls and corners, and roof bands are the means to tie the roof structure to the wall structure. Ring-beams can be made from either timber or reinforced concrete, concrete being preferred in areas where timber is in short supply or forests are protected. Roofs may be flat or pitched, but they should be light in weight, well-connected to the walls through a ring-beam, and cross-braced in the plane of the roof. Construction details for ring-beams and for roof and floor construction are given in the WHE Tutorial (Bothara and Brzev 2011). The tutorial also gives good recommendations of the choice and preparation of building stone, mortars and concrete for ring-beams, and for the construction of foundations.

In the 1980s, the authors' team worked with the Turkish Government's Earthquake Research Institute in Ankara on ways to improve the traditional stone masonry dwellings of eastern Turkey, huge numbers of which had been destroyed in several earthquakes of the preceding decade, with considerable loss of life (Spence and Coburn 1987). An Impulse Table was constructed on which full-sized replicas of stone masonry dwellings were built by masons from the affected areas, one type without ring beams, one type with traditional timber ring-beams (*hatil* in Turkish) and one type with reinforced concrete ring-beams. The ring-beams were placed at roof level, at lintel level and at mid-height of the door opening. The Impulse Table was able to demonstrate the comparative performance of the three prototypes under progressively increasing levels and durations of shaking. The unreinforced test building failed under low ground shaking, but both the test pieces with ring-beams performed well (Figure 5.4). The results were used to demonstrate to the local masons the benefits of the strengthening methods adopted.

Figure 5.4   Impulse Table testing of stone masonry houses in Eastern Turkey: test house with triple timber hatils. *Source:* Spence and Coburn (1987).

For retrofitting existing stone masonry buildings, enhancing or creating integrity is again the main objective. A big improvement of seismic performance can be achieved by tying external walls together by using steel ties passing through the building, under floor or roof level, and with anchor plates at each end. This is a very ancient technique much used in southern Europe, the effectiveness of which has been demonstrated in many earthquakes, and studied experimentally in Slovenia (Tomazevic 1999). Alternatively, steel ties or reinforced concrete bandages can be used to wrap around the entire building at the level of floors and just below roof level. With rather more intervention, reinforced concrete ring-beams can alternatively be introduced at the tops of the walls to tie walls and roof together; and existing floors and roofs can be tied to the walls by the addition of metal ties. However, added reinforced concrete ring-beams were found to be a source of weakness following several Italian earthquakes (EEFIT 2016a).

Poorly interconnected external and internal masonry wythes can be stitched together with reinforced concrete stitches anchored in each face of the wall. Or jacketing can be used, which involves covering both internal and external faces of a wall with a thin layer of reinforced mortar containing welded wire mesh or polymeric materials with the two layers tied together through the wall. Details are given in World Housing Encyclopaedia's Stone Masonry Tutorial (Bothara and Brzev 2011). Some of these techniques are likely to be unaffordable for non-engineered buildings, but can be valuable for historic urban centres where maintaining the architectural heritage is of paramount importance. For improving non-engineered buildings, it is likely to be more effective to concentrate on improving standards of new buildings rather than retrofitting older ones.

5.2.3 Unreinforced Brick and Block Masonry Construction

Fired clay brick masonry has been the principal walling material used for urban buildings for many centuries throughout much of Europe, North America, India and China. The basic building material, fired clay brick, is made from abundant soils and clay beds, the firing process is relatively simple, and the resulting building material, brick masonry, is suitable for a wide variety of building types, single up to many storeys, and is adaptable to many building plan forms and uses, homes, schools, hospitals and leisure buildings. Brick masonry is cheap and, if well-made, very durable (D'Ayala 2004). But other forms of unreinforced masonry have become common. In some parts of the world, concrete block masonry is today the predominant technology because it is cheaper or clay soils are not available; unit masonry walling can also be made from stabilised soil blocks (Spence 1971), or dressed stone blocks where the materials are available and easier to use. The various forms of unreinforced masonry (URM) today constitute over 50% of the residential building stock in a number of the world's largest earthquake-prone countries including India, China and Indonesia, and over one-third in many more countries, including Turkey, Colombia, Ecuador, Italy and Australia.

Masonry walls are built by laying the bricks or blocks in a mortar made from sand, cement or lime and water. The masonry units are laid in parallel in courses, with the joints staggered to create integrity to the wall; adjacent wall elements can readily be bonded to ensure continuity of strength. A variety of styles of bonding are used according to local traditions (Coburn et al. 1995; Spence and Cook 1983). Floors and roofs of timber rest on the walls usually using a timber wall-plate, or sometimes joist hangers (Figure 5.5).

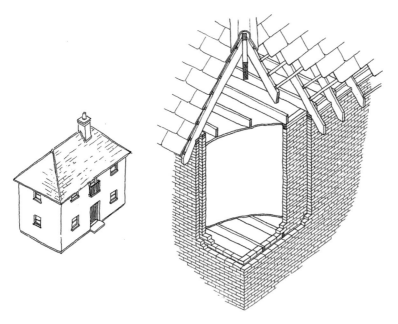

Figure 5.5 Traditional unreinforced brick masonry.

If, as in more modern construction, floors and roofs are reinforced concrete slabs, these rest directly on the inner leaf of the walls. Cavity walling, in which inner and outer leaves are separated by an insulated cavity to provide better insulation, is today common in temperate or cold climates, and metallic cavity wall ties are used to connect the two leaves.

Unreinforced masonry buildings are likely to be damaged in moderate-to-strong earthquakes. In moderate earthquakes up to EMS intensity VI, URM buildings generally resist earthquakes with slight damage, but at intensity EMS VII or more damage occurs through cracking of walls or overturning of wall sections not well-tied to roofs and floors, and at higher intensities (EMS VIII or above) complete collapse can occur through the failure of the bearing wall. The typical damage states are illustrated in Figure 5.6. Chimneys, parapets and gable walls are particularly liable to collapse through overturning, with the potential to cause casualties to people outside the building (Figure 5.6). The walls of URM buildings with large window openings are particularly vulnerable as this can leave narrow piers of masonry between openings exposed to higher stresses. X-shaped shear cracking patterns can develop starting at the corners of openings.

As an extreme example, the collapse of unreinforced brick masonry buildings was responsible for most of the 250 000 deaths reported in the 1976 Tangshan earthquake in China, and also for many of the deaths in the Mw6.0 Skopje (North Macedonia) earthquake of 1963. Many unreinforced brick masonry buildings also collapsed in the Mw7.7 2001 Bhuj earthquake in India and in the Mw7.9 2008 Wenchuan earthquake in China (Figure 2.7), and there was immense damage to older unreinforced masonry in the Mw6.1 2011 Christchurch earthquake in New Zealand (Figure 5.7). Even minor earthquakes, such as the Mw5.3 2020 Zagreb earthquake in Croatia, can cause significant damage to unreinforced brick masonry buildings.

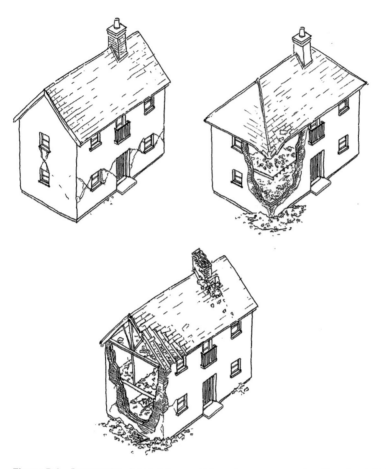

Figure 5.6 Progressive damage in unreinforced masonry construction; top left: X-cracking, top right: corner failure, below: gable wall separation and overturning.

Figure 5.7 Wall collapse in masonry building with timber upper floor and roof, Christchurch earthquake, New Zealand 2011. *Source:* Author owned image.

The essential requirements for earthquake-resistant design and construction for brick and block masonry are (Coburn et al. 1995):

- Robust and regular building form
- Firm foundations
- Good quality materials
- Strong walls
- Distributed openings

The plan of the building should be arranged so that it is regular and symmetrical, with closely spaced walls running in both directions; damage is likely to be worse if the plan is elongated or irregular; and upper storeys should not have overhangs or setbacks. Walls need to be built on a substantial foundation made from a brick, block or stone plinth if the subsoil is good, or reinforced concrete if the subsoil is poor. The quality of the brick or block materials and of the mortar is of primary importance. Materials should be tested according to available standards. Walls need to be well-bonded in a manner appropriate to the type of masonry used, with particular attention to ensuring bonding at wall-to-wall junctions. For cavity walls, there need to be regularly spaced wall ties made of non-corrosive material to avoid separation and collapse of the outer layer. And openings need to be relatively small, well-spaced and located away from the corners. Various countries' manuals and codes of practice for different types of brick/block masonry give specific detailed and dimensional recommendations (CEN 2004; Government of Nepal 1994b).

A variety of methods are available to retrofit unreinforced masonry buildings. The techniques used for retrofitting stone masonry buildings (Section 5.2.2) can also be used for brick and block masonry, including adding concrete ring-beams, tying walls by means of external wall ties or bandages, and jacketing. In addition, local repairs to damaged walls can be done through repointing – replacing poor or damaged mortar in the wall face with better quality mortar, injection of cement paste into existing cracks, or by stitching using added local reinforcing bars, which are grouted into the wall. These techniques are described in more detail by Dina D'Ayala (2004), and in Chapter 21 of FEMA 547 (FEMA 2006). The government of New Zealand has prepared comprehensive guidelines for the assessment of existing buildings of unreinforced masonry, to enable the need for retrofit to be determined (Government of New Zealand 2017).

5.2.4 Confined and Reinforced Masonry Construction

Confined masonry and reinforced masonry are construction techniques developed over the last 100 years to improve resistance of masonry buildings to lateral loads. Confined masonry has become popular over the last 40–50 years in many of the high-risk earthquake areas, where masonry has been the traditional form of construction, not only in Mexico and Chile and other South American countries but also in earthquake-risk countries in Europe, notably Slovenia, Croatia, Serbia and Romania. In these European countries, construction is usually under the direction of structural engineers and is governed by local codes of practice, but there are other countries where a form of confined masonry is used for informal owner-built housing. In Chile, confined masonry is estimated to constitute about 30% of the country's building stock (Alvarez Velasquez et al. 2016), while in Mexico, confined masonry is now estimated to constitute as much as 60% of the national housing stock (Rodriguez 2004).

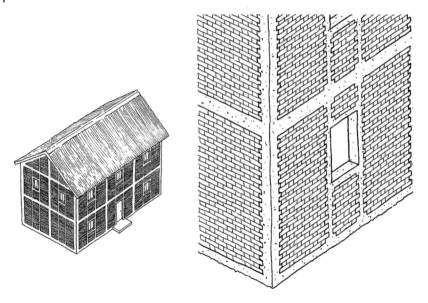

Figure 5.8 Confined masonry construction, showing typical locations of confining elements.

In *confined masonry*, unreinforced masonry walls are confined by vertical reinforced concrete tie-columns and horizontal reinforced concrete tie-beams. The tie-columns are not designed to carry the vertical loads, but act together with the tie-beams to provide confinement to the masonry walls in their plane, improving their resistance to lateral in-plane loads, and also preventing out-of-plane failure. Vertical elements are needed at a minimum at the corners of a building, at the free ends of walls, and around openings (Booth 2014) (Figure 5.8). A variety of types and sizes of masonry units can be used, and confined masonry is used for buildings up to four storeys, for example in Chile (Rodriguez 2004).

A very important feature of confined masonry construction is that the tie-columns and beams are cast in place *after* each level of masonry construction has been completed, to ensure a strong bond between the concrete and the masonry (Figure 5.9). Floors are generally formed from reinforced concrete slabs with hollow-block infills, and roofs may be of reinforced concrete construction, or sloped timber roof beams are used in conjunction with sheet roofs.

Confined masonry buildings have been affected by earthquakes in Mexico (Mw7.1 Puebla 2017) and Chile (Mw8 in 1985 and Mw8.8 in 2010). In both countries, it is reported (Rodriguez 2004) that damage was light (see Chapter 8). In the 2010 Chile earthquake, in spite of its large magnitude, only two confined masonry buildings collapsed. In Chile, observed damage patterns, largely affecting buildings built with poor materials or insufficient tie-columns, included the following:

- Shear cracks in walls propagating into tie columns.
- Horizontal cracks at the joints between masonry walls and reinforced concrete floors or foundations.
- Crushing of concrete at the joints between vertical tie-columns and horizontal tie-beams.

Seismic performance was better in buildings which had a regular plan and sufficient amount of walls in each horizontal direction (Astroza et al. 2012).

Figure 5.9 Confined masonry construction in Chile: detail of wall-column interface prior to column casting. *Source:* World Housing Encyclopaedia. Reproduced with permission of EERI.

However, in some areas, buildings apparently of confined masonry have been found to be very poorly constructed, using substandard materials. These were often built with a gap between top of the masonry wall and the underside of the concrete tie-beam, so that vertical loads cannot be transmitted into the walls, and lateral loads need to be transmitted though short columns. There was much damage to such 'pseudo-confined-masonry' construction in the Mw7 2010 Haiti earthquake (EEFIT 2010a). However, this building type should perhaps be described a non-engineered reinforced concrete frame, rather than confined masonry.

In *reinforced masonry*, steel reinforcement is placed within masonry walls. Vertical reinforcement is threaded into the cavities in hollow concrete blocks or hollow-clay blocks, spaced at regular intervals along the length of the walls, which are then infilled with cementitious grout to form a continuous vertical reinforcing element (Figures 5.10 and 5.11). Or if solid masonry units (for example clay bricks) are used, the bonding of the walls is arranged to provide a cavity within the wall into which reinforcement can be placed. Horizontal reinforcement is placed in the mortar layer at regular intervals up the height of the wall, and above and below openings.

Spacing and size of the reinforcement, and details for connections to floors, roofs and foundations are given in national buildings codes, as in Colombia (Government of Colombia 2011), where this form of construction is encouraged, and economically supported by the governments of large cities for low-income families (Hackmayer et al. 2002) (Figure 5.12).

Reinforced masonry has been practised in the United States, Canada and New Zealand since the 1930s, and there is a growing trend in Latin American countries, Mexico, Chile and Colombia.

One of the first recorded examples of the use of reinforced masonry was in the housing built for the Military Cantonment of Quetta in Pakistan, where the brick masonry was reinforced by steel reinforcing bars, vertically and horizontally, using the so-called 'Quetta

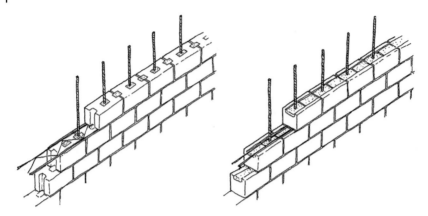

Figure 5.10 Reinforced masonry construction. Typical reinforcing details.

Figure 5.11 Reinforced masonry using concrete block masonry in Canada. *Source:* Bill McEwan. Reproduced with permission.

bond'. Following an earlier earthquake in 1931, this system had already been developed and used for army housing before the devastating M 7.5 1935 earthquake, and a report on that event (Jackson 1960), related that while ordinary housing was completely destroyed, the Quetta bond construction of the military housing was undamaged. The city of Quetta adopted reinforced masonry in its 1940 building code and a visit by our research group in 1980 found the method is still in use (Spence 1983).

Figure 5.12   Reinforced masonry construction for urban housing in Colombia, 2011. *Source:* Photo: World Housing Encyclopaedia. Reproduced with permission of EERI.

The performance of reinforced masonry buildings in recent earthquakes has been relatively good. Some were damaged in the Mw6.1 2011 Christchurch in New Zealand and the Mw8.8 2010 Maule earthquake in Chile, but they did not collapse (Brzev 2020).

5.3   Reinforced Concrete Construction

Reinforced concrete is today the most common structural material used in the construction of urban multi-family housing worldwide and constitutes a growing proportion of the world's total building stock. Its use has been widespread since the 1950s for urban housing, as well as for commercial buildings (hotels, offices), and in many cities for mixed-use buildings with commercial use or car parking located on the ground floor and housing above. Its popularity derives from its cheapness, from the ability of reinforced concrete to be used for multi-storey construction, enabling greater density of built space in congested urban areas with high land prices, and also from the better opportunities for space planning within a building, allowing for larger openings and greater flexibility in placing partition walls than are possible in masonry buildings.

In most European countries, the proportion of the urban housing stock which is built using some form of reinforced concrete ranges from 25% to over 75%. It is also used very widely in the urban areas of the developing world, comprising for example approximately 75% of the urban building stock in Turkey (Figure 3.10) and about 60% in Colombia (Jaiswal and Wald 2014).

The basic components of reinforced concrete frame construction are vertical columns and horizontal beams forming a three-dimensional frame, with reinforced concrete slabs spanning between the beams to form the floors and roofs. All these elements are reinforced with steel reinforcing rods which are arranged to provide continuity of reinforcement between columns, beams and slabs, creating a monolithic construction. If properly designed, such a structural arrangement is capable of carrying both gravitational loads and the lateral loads arising from wind or earthquake loading.

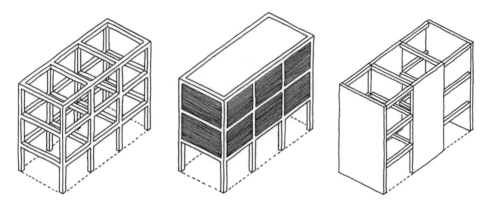

Figure 5.13 Forms of reinforced construction. From left: concrete frame, infilled concrete frame with soft storey, concrete shear wall construction.

However, early reinforced concrete frame buildings were often found vulnerable to earthquakes because of lack of ductility as well as strength, deriving from poor materials, poor arrangement of the structural members, or poor detailing of the reinforcement. It has been observed that buildings using reinforced concrete walls, oriented in both longitudinal and transverse directions, rather than comparatively slender columns as the vertical elements of the structure, are more robust in their earthquake response. This markedly reduces their vulnerability to earthquake ground shaking. So it is usual to make a primary sub-division of reinforced concrete structures between reinforced *concrete frame* and *reinforced concrete shear-wall* structures (Figure 5.13). In many countries, a combination of shear wall and frame, a so-called 'dual system', is widely used.

Another distinction, which can have important implications for earthquake vulnerability, is between *cast-in-place* concrete construction, in which the concrete is cast on site, allowing direct continuity of reinforcement between one cast section and the adjacent one, and *precast* concrete construction, in which the individual elements are cast off-site and joined together on site by welding or connecting projecting reinforcing bars with local cast-in-place concrete 'stitches'. Both cast-in-place and precast construction are used for shear-wall as well as for frame structures.

5.3.1 Reinforced Concrete Frame Construction

The structural system of a reinforced concrete frame structure consists of interconnected beams and columns, with floor and roof slabs spanning onto the beams. Joints between beams and columns are moment resisting, i.e. designed to carry both bending and shear force. Walls are generally formed by non-structural infill materials, such as lightweight concrete blocks, solid clay bricks or hollow fired-clay blocks or other perforated bricks, which are not designed to provide lateral stiffening of the frame.

Reinforced concrete frame structures can be of any height from single storey to high-rise structures. In many countries, the most common height is three to seven storeys. Reinforced concrete frame structures often have the ground floor left partly or totally free of walls, in order to facilitate commercial or car parking use. The ground floor storey height may also be greater than that of upper floors, making the ground floor much more flexible than upper floors. Such structures are often referred to as *soft-storey* structures (Figure 5.13).

Reinforced concrete frame buildings have been tested in many earthquakes in the last 50 years, notably in the Mw7.6 1999 Kocaeli earthquake in Turkey, in the Mw7.7 2001 Bhuj earthquake in India, the Mw7.9 2008 Wenchuan earthquake in China and the Mw7.8 2015 Gorkha earthquake in Nepal, and often performed very badly. Very common sources of earthquake damage have been (Yakut 2004) the following:

- Shear failures and concrete crushing failures in columns, leading sometimes to failure of the column under gravity loads and local or total collapse of the building.
- Soft-storey effects: as explained above, the use of an open and often taller ground floor for commercial or parking causes a lack of stiffness at that floor which leads to excessive deformation resulting in heavy damage or total building collapse.
- Short-column effects: restraining of columns by walls over only part of their height leading to high stresses and local failure of the column.
- Torsional failures due to the use of irregular plans or changes in plan from one floor to another.
- Poor detailing of reinforcement: inadequate laps of bars, insufficient lateral hoop reinforcement.
- Substandard quality of concrete.

Some of these types of failure are illustrated in Figure 6.17 and a particular example of soft storey failure is shown in Figure 5.14.

Figure 5.14   Soft-storey collapse of reinforced concrete building, Kocaeli earthquake Turkey, 1999. *Source:* EEFIT. Reproduced with permission.

Codes of practice developed over the last 50 years, if followed, should prevent these types of failure. Codes developed since the 1970s have emphasised the need for ductility so that buildings can deform extensively during an earthquake without loss of strength (see Chapter 7). But in many of the areas where these buildings have performed poorly, this was either because the buildings were built before current codes were introduced, or the existing code was not followed, due to inadequate code enforcement in design, or because of lack of supervision on site. Such non-engineered reinforced concrete buildings are today very numerous in rapidly developing cities in earthquake-risk zones in Asia (Jain 2016) and South America (Cardona 2019).

Techniques for retrofitting existing reinforced concrete buildings to enhance their earthquake resistance have been developed and tested in laboratories. They include the following:

- strengthening existing beams and columns, by jacketing them with concrete reinforced with steel mesh or fibre wrap overlays;
- adding new infill walls in such a way as to work in conjunction with existing beams and columns in carrying lateral loads.

These techniques are described in detail by Booth (2014) and FEMA (2006). They have been tested in the laboratory (Jong et al. 2017), and they have been applied to a number of buildings, e.g. in in Turkey and Romania. To date however, little evidence of their performance in subsequent earthquakes is yet available.

5.3.2 Reinforced Concrete Shear-Wall Construction

The structural system of a reinforced concrete shear-wall building is similar to that of a reinforced concrete frame building except that the vertical elements are walls (substantially longer than their width), rather than columns. The walls are oriented in two perpendicular directions, carry both vertical and lateral loads, and are continuous throughout the building's height. They are continuously reinforced, normally in both faces of the wall, and special attention is given to the reinforcement of the wall end zones, and adjacent to door or window openings. The slabs span directly onto the walls with no beams.

Reinforced concrete shear-wall buildings may be either cast-in-place or constructed using precast concrete elements, using stitching or welding to create reinforcement continuity. In a number of countries, especially Russia and former USSR Republics, many thousands of housing units have been constructed using such precast concrete wall systems.

Shear-wall structures have generally performed very well when tested in earthquakes in many different countries over the last 50 years, including the Chile earthquakes of 1985 and 2010, where this type of building was widely used for high-rise housing, Figure 3.24 (Moroni and Gomez 2011). Shear-wall structures using a cast-in-place tunnel-form system performed very well in the Mw7.6 1999 Kocaeli earthquake in Turkey, when many concrete-frame buildings collapsed (Moroni 2004). Large panel precast concrete building systems used in Eastern European and Central Asian countries have also performed well in a number of earthquakes such as the 6.8 1988 Spitak Armenia earthquake, and the Mw7.5 1977 and Mw7 1990 earthquakes in Romania (Brzev 2004). In the Mw8.8 2010 Maule earthquake in Chile, the EFFIT team reported that in Concepción, the worst-affected city,

Figure 5.15   Collapse of reinforced concrete shear wall structure, Conception, 2010 Maule earthquake. *Source:* EEFIT. Reproduced with permission.

modern construction using shear-wall structures generally performed well. However, there were some significant high-profile failures in very recently constructed buildings, many of these being no more than five years old (EEFIT 2010b). The most prominent of these was the 15-storey Torre Alto Rio building, which totally collapsed (Figure 5.15). Subsequent analysis based on local ground motion recordings indicated that this building was subjected to very strong ground shaking, placing a displacement demand of two to three times that designed for, lasting for two minutes; the T-shaped cross-section of key shear walls was also shown to have probably led to damaging strains in the concrete. The analysis suggested the need for changes in design codes internationally (Tuna and Wallace 2019).

Because they have generally behaved well in earthquakes, there has been relatively little work on retrofitting reinforced concrete shear-wall structures. Some structures have been retrofitted in Romania (Moroni 2004), either to increase the resistance of walls of inadequate strength (using jacketing techniques), or to repair cracks in earthquake-damaged buildings using injected epoxy resin. Some damaged buildings were also retrofitted after the 2010 Maule earthquake in Chile (Sherstobitoff et al. 2012).

5.4   Timber Frame Construction

Timber construction is commonly used for single-family houses in many countries of the world. In western USA, Canada, Japan and New Zealand, timber stud-wall construction is today used for the vast majority of single-family houses, and in many other countries, it forms a significant proportion of new building stock. In the Western USA, it accounts for

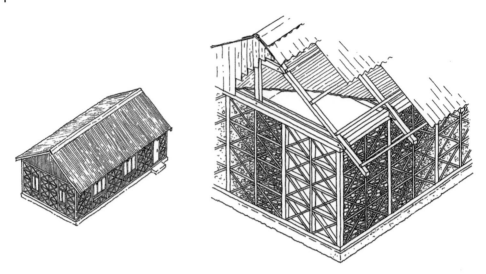

Figure 5.16 Modern Dhajji Dewari construction.

about 98% of existing and new houses constructed (Arnold 2004). In some cities timber-frame buildings are also used in multi-family housing. In Europe, timber-frame housing is used, but constitutes a small proportion (less than 5%) of the housing stock, except in Romania, where timber post and beam construction accounts for 14% of the existing single-family housing stock. Where it is very widely used, the primary reasons for the popularity of timber-frame housing are the easy availability of suitable timber and its affordability. But its popularity also comes from the experience of past earthquakes which have demonstrated the superior performance of timber relative to other (particularly unreinforced masonry) forms of construction. In some areas, regulation prohibits the use of unreinforced masonry.

Timber construction has also been used traditionally in many countries, using a variety of methods, including post and beam construction, half-timbered or wattle and daub construction in Northern Europe, *Shinkabe/Okabe* construction in Japan (Figure 3.15), *Himiş* and *Bağdadi* construction in Turkey, *Dhajji Dewari* construction in Kashmir, India (Figure 5.16). It has also been used for lightweight timber frame structures in tropical regions such as Indonesia (Figure 3.2) or South India, and pole structures covered with thatch or mud-plastered walls of bamboo or reeds (*Bajareque* in Central and South America), and traditional construction in many parts of Africa. In most of these countries, such buildings are still built, or form a significant part of the existing building stock. Several of these forms of construction, too, have often demonstrated a superior performance in earthquakes compared with local forms of unreinforced masonry or reinforced concrete. However, in a number of countries (for example in India) where timber has been used traditionally for housing, its use in building is now prohibited to protect the remaining forests.

Timber is also, more rarely, used in other ways, such as log construction,(still found in Scandinavia and Northern Europe, Switzerland, the Balkan countries and the Russian Federation) and wood panel construction (also in the Russian Federation) (Arnold 2004). And wherever timber is available, it is also used for non-residential buildings such as barns and warehouses using a variety of building techniques. However, very rarely has timber been used for buildings greater than three or four storeys in height, partly because of structural limitations and also because of fire risk. But timber is a low-energy and ecologically sustainable material, and, as engineered timber has significantly increased strength, its use for larger and taller buildings may well increase substantially in the coming years (Ramage et al. 2017).

This section will focus on stud-wall construction, as this is the most widespread current form of timber construction. It is predominantly used for one or two-storey single-family housing, but has been in some cases also used for multi-family housing in two-storey or three-storey terrace blocks.

5.4.1 Timber Stud-Wall Construction

A typical timber stud-wall building (Figure 5.17) consists of a reinforced concrete strip foundation, on which the timber ground floor is constructed and anchored to the foundation; the ground floor consists of a platform of softwood timber joists covered with a deck of plywood or other timber-based board; on this platform exterior and interior ground floor walls are built, consisting of a timber sill plate and regularly spaced (30–60 cm centres) vertical timber studs of a suitable width. Board or panel sheathing is nailed to the studs on the outside of the building (Arnold 2002). The studs are capped with a double header plate and the first floor is built on this, in turn acting as a platform for the first floor walls, on which a roof of timber rafters or prefabricated trusses is built, and covered with tiles or roof

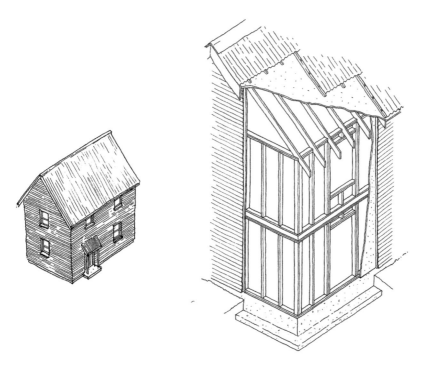

Figure 5.17 Timber stud-wall construction – general form and detail of corner construction.

sheets. An external cladding of brickwork or of horizontal timber boarding is also sometimes added.

Local variants of the above description are that in some cases, a basement is built in reinforced concrete, to which the ground floor assembly is bolted. In other cases the ground floor is elevated above the foundation using short stub or 'cripple' walls. In Japan, posts of square cross section about 105×105 mm are used at wider spacing (typically at 1.8–2 m), and lateral resistance is sometimes provided by internal diagonal bracing members rather than plywood sheathing.

In the USA, Japan and New Zealand most timber frame construction is covered by locally applicable standards which define minimum requirements for the material properties, sizing and spacing of studs, the size of plywood or other stiffening panels and the means of connecting the members, including anchorage to the foundations and the size of openings. Adopting these standards ensures sufficient resistance to expected earthquake forces in the region and that the building has adequate redundancy to sustain very limited damage in expected earthquakes, without the need for additional structural design calculations. Nevertheless, some damage to timber frame buildings has occurred in recent earthquakes as illustrated in Figure 5.18 and described below.

In California, the Mw6.7 1994 Northridge earthquake caused severe ground shaking to many thousands of one- and two-storey conventionally designed residential buildings, with little damage caused beyond loss of brick chimneys, cracked plaster and broken glass (EEFIT 1994), and some failure of cripple walls in older buildings. However, some two and three storey buildings, particularly those with open tuck-under parking, suffered extensive damage, and in a number of these cases the ground floor partially or totally collapsed.

In the Mw6.1 Christchurch, New Zealand earthquake of 2011, the majority of buildings affected were also one and two-storey timber-frame houses. For these structures, damage was primarily driven by ground failure around or beneath the structure, specifically differential movement of separate structures and fracture of concrete block foundations due to lateral spreading and settling of foundations in areas subject to liquefaction (EEFIT 2011). Commonly observed damage also included damage to masonry components of the building: falling chimney stacks, damage or collapse of masonry facades or to external chimney breasts; and roof tile damage. Several instances of soft storey failure in a timber building were also observed (Figure 5.19) (EEFIT 2011).

After the Mw7.0 2016 Kumamoto earthquake in Japan, a building damage survey was carried out by EEFIT (2016b) in Mashiki, a town close to the main activated fault, where ground motion was very severe, and considerably exceeded that for which modern buildings are designed. There was serious damage not only to older buildings, designed before the 1981 building code, but also many cases of failure in more modern buildings. The most catastrophic failures were complete collapse, involving disintegration of timber walls and roofs, affecting the older buildings. Among the buildings built to the 1981 code, there were cases of substantial lateral drift or even collapse of the bottom storey in two-storey houses in buildings, where the walls were not strengthened by sheathing boards, but by diagonal struts which had failed (Figure 5.20). A survey showed that about 50% of timber frame buildings in Mashiki suffered serious damage (EMS Grade 3) or worse (EEFIT 2016b).

Figure 5.18 Typical forms of damage to timber stud-wall structures. From top left: undamaged, wall distortion, lower-storey wall disintegration, and falling masonry elements.

Figure 5.19 Damaged timber-framed building in Christchurch, New Zealand, in the 2011 earthquake. *Source:* Buchanan 2011. Reproduced with permission.

Figure 5.20 Damaged recently built timber-frame building in Mashiki Town, in the 2016 Kumamoto earthquake. *Source:* EEFIT. Reproduced with permission.

Arnold (2004), reviewing global performance of timber frame houses, reported that seismic deficiencies found in older timber-frame houses included inadequate connection of the building to its foundation, resulting in the building slipping off its foundation. A further deficiency found particularly in the USA is the use of unbraced cripple walls, which can overturn or collapse in an earthquake causing severe damage to the main structure. Older buildings may also lack adequate shear resistance through sheathing boards which are either inadequate in size, not continuous, or insufficiently connected to the vertical studs. In Japan particularly, heavy roofs, coupled with inadequate bracing, have often been the cause of collapse of significant numbers of older timber frame houses, as occurred in the 1995 Kobe earthquake. In older buildings, the timber may be of poor quality, may suffer from rot, fungal attack or insect attack, especially if the timber is not treated, and metallic connections may suffer from corrosion. All of these factors will significantly increase the vulnerability of the building.

Some retrofit of particularly vulnerable timber-frame buildings has been carried out both in Japan and in the USA. In Japan, a number of different retrofit techniques have been adopted. Metal connectors have been added to strengthen inadequate framing-member connections and wall-to-foundation connections. For buildings with inadequate lateral bracing (the main cause of failure identified in the Kumamoto earthquake), additional timber diagonal braces, or plywood sheathing has been added. And heavyweight roofs have been replaced by lighter roofs to reduce the lateral forces on the wall structure. In the USA, the main technique has been the bracing of cripple walls at foundation level (Arnold 2002).

5.5 Steel Frame Construction

Steel is in many ways an excellent material for construction in earthquake areas, because of its high strength combined with ductility. It has been used, largely successfully, in Japan, the United States and other countries for many years. The USGS PAGER Global Building Stock Inventory also records significant use in urban areas in several other countries, notably Chile, Russia, Indonesia and Iran (Jaiswal and Wald 2014). Its economic use is largely not only for high-rise buildings, particularly for commercial use (e.g. office buildings) but also for low-rise commercial (retail and warehouse) and industrial buildings, where the building layout demands long-span roofs. However, recent earthquakes have identified a number of problems with the connections in steel frame buildings built before the 1990s, which subsequent codes of practice have aimed to overcome.

Among steel frame structures, it is useful to distinguish moment-resisting frames from braced frames which resist lateral load in different ways. Moment-resisting frames have no diagonal (bracing) members, the resistance to lateral load being provided by the rigidity of the beam-column joints, which are either bolted or welded. In braced-frame structures, diagonal steel members are used to provide the resistance to lateral loads, and a variety of bracing systems can be used, depending on the design requirements of the building (Figure 5.21). Some older steel frame buildings use infill masonry walls to provide lateral resistance, though this type of structure has a poor record in earthquakes.

The floor and roof act as either flexible or rigid diaphragms and are usually made of cast-in-place concrete slabs, precast slabs or metal deck with concrete fill supported on steel beams with open web joists, or steel trusses. For moment-resisting or braced frames, the structure is often concealed by exterior non-structural walls, which can be of almost any material (curtain walls, brick masonry, precast concrete panels or glass).

Properly detailed moment-resisting frames are capable of dissipating large amounts of energy under deformation and therefore perform well in earthquakes; however, such structures are typically more flexible than braced frame structures, which can result in large inter-storey displacements and consequent non-structural damage. Incomplete braced frames, i.e. structures with either braces in only one direction or bays with missing braces, can significantly increase a structure's vulnerability. This is due to large inter-storey drifts in unbraced bays leading to possible instability and collapse.

A further type of steel frame structure used mainly for commercial or light-industrial buildings is a light steel frame (Figure 5.22). In some countries, this structure type has also been used to provide low-cost housing that can be rapidly constructed, for example in post-disaster emergency situations. Light steel frames are usually single-storey structures with relatively long roof spans. The structure consists of moment-resisting frames in one direction, which are joined by bracing rods or bars or lightweight cladding in the

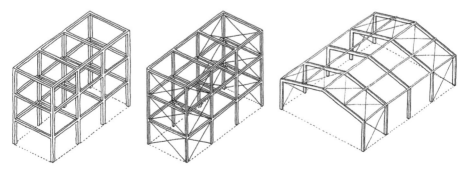

Figure 5.21 Forms of steel frame construction. From left: moment frame, braced frame and light steel frame.

Figure 5.22 Collapsed 21-storey steel-frame Pino Suarez building, Mexico City, in 1985 earthquake. *Source:* EEFIT. Reproduced with permission.

direction perpendicular to the frame. The frames are built in segments and assembled on site with bolted joints. The roof and walls are usually lightweight panels such as corrugated metal, especially in the non-residential sector. The frames are often designed for maximum efficiency with tapered beam and column sections.

In spite of one particular high-profile collapse, the 21-storey Pino Suarez Building in the Mw8 Mexico City earthquake in 1985 (Figure 5.22), steel frame buildings were regarded as very safe in earthquakes until the early 1990s (Figure 5.23). This may be because they are mostly built in large urban centres and had hardly been tested under heavy ground shaking until that time (Hamburger 2003). In the Mw6.7 1994 Northridge earthquake in California, which tested many high-rise braced and moment frame buildings under heavy ground shaking, no steel building collapsed, but many were found subsequently to have developed very serious cracks in the welded beam-to-column connections, which required extensive repairs. And in the following year, in the Mw6.9 Kobe earthquake in Japan in 1995, a large number of steel-framed buildings collapsed, with the failures often initiated at connections (Figure 5.24). Extensive research followed these failures, and new code revisions were put in place which should have made more recently constructed steel buildings safer. But these

Figure 5.23 Latino American Tower, Mexico City, in 2015. This landmark steel-frame building survived the 1985 earthquake undamaged. *Source:* Photo https://en.wikipedia.org/wiki/Torre_Latinoamericana.

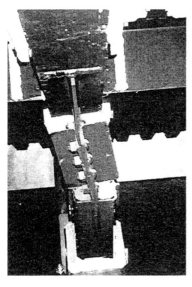

Figure 5.24 Details of failed connections in steel-framed buildings, 1995 Kobe earthquake. *Source:* EEFIT. Reproduced with permission.

recent buildings have not yet been subjected to loads near to or exceeding their design level (Booth 2014; Hamburger 2003).

By comparison with the Kobe and Northridge experience, high-rise steel frame buildings in Chile, designed to Chile's earthquake code, are reported to have survived heavy ground shaking in several potentially damaging earthquakes (in 1960, 1985 and 2010) with no serious damage (Arze 2002).

In Iran, steel frame buildings have been used for urban construction for many decades, and a system of construction has been developed and widely used for apartment buildings up to nine storeys, in which a steel frame is infilled with masonry walls, and floors and roofs are also composed of masonry jack-arch construction, spanning between steel joists. The frame is moment-resisting in one direction, but braced in the perpendicular direction (Alimoradi 2002). Many buildings of this type were seriously damaged or collapsed in the Mw7.4 Manjil earthquake in 1990 not only due partly to failures of welded connections, but also due to the weak behaviour of heavy but flexible floors and roofs. These deficiencies have been addressed in a more recent code of practice, but older steel buildings have continued to be damaged in subsequent earthquakes.

Retrofitting strategies for existing steel buildings which have inadequate resistance can be through the addition of cross-bracing to the existing frame, thus reducing the amount of displacement experienced in earthquakes and helping to protect non-structural elements. Or an additional braced frame can be placed alongside the existing frame, leaving the existing frame to carry only gravity loads.

For cases where the beam-column welded connections are suspect or inadequate, techniques have been developed in the USA to strengthen the joints with additional welded plates so that failure has to occur in the beams, limiting the danger of collapse. Methods are discussed by Booth (2014), and more detail is given in FEMA 547 (FEMA 2006).

5.6 Comparing the Vulnerability of Different Construction Types

A way to measure the relative earthquake-resistance of different construction types is to define and measure their *vulnerability*. This chapter will conclude with a brief overview of vulnerability, to enable a quantitative comparison to be made between the earthquake performance of different construction types.

In relation to earthquake ground shaking, vulnerability can be defined as 'the degree of loss to a given element at risk resulting from a given level of earthquake ground shaking' (Coburn and Spence 2002). Ground shaking may be measured by scales of Macroseismic Intensity which are based partly on damage observations. Or ground shaking may be measured in terms of physical parameters of ground motion such as peak ground acceleration (PGA) or spectral acceleration (SA).

Loss can be measured in terms of the cost to repair the damage caused by the earthquake, or as a proportion of buildings of a particular type suffering a particular level of damage. Because repair costs vary greatly according to local factors, we will here use a set of defined damage levels. It is common, using the definitions of the European Macroseismic Scale (EMS-98) (Grünthal et al. 1998) to define six different damage levels ranging from D0 (no damage) to D5 (complete collapse). Table 5.2 defines the EMS-98 set of damage levels we

Table 5.2 Brief definitions of damage states for masonry and reinforced concrete buildings.

Damage level	Damage level description	Definition for load-bearing masonry buildings	Definition for reinforced concrete frame buildings
D0	Undamaged	No visible damage	No visible damage
D1	Slight damage	Hairline cracks	Infill panels damaged
D2	Moderate damage	Cracks 5–20 mm	Cracks <10 mm in structure
D3	Heavy damage	Cracks >20 mm or wall material dislodged	Heavy damage to structural members, loss of concrete
D4	Partial destruction	Complete collapse of individual wall or roof support	Complete collapse of individual structural member or major deflection to frame
D5	Collapse	More than one wall or more than half of roof collapsed	Failure of structural members allowing fall of roof or floor

Source: Based on Grunthal et al. 1998.

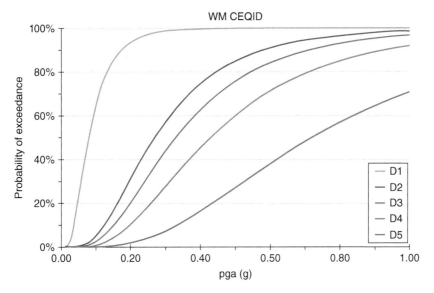

Figure 5.25 Mean probability of a building of weak masonry exceeding each of the five levels of damage (D1–D5) defined in Table 5.2, as a function of peak ground acceleration (PGA). *Source:* Based on damage data assembled in the CEQID database.

will use here and gives the corresponding brief definitions of damage, for load-bearing masonry and for reinforced concrete buildings.

For a particular class of building, it is possible to plot the average vulnerability, or *fragility*, against the level of ground shaking, using purely observational data, as shown in Figure 5.25.

The damage data points in Figure 5.25 are derived from damage surveys in several hundred locations in 79 earthquakes worldwide, while the ground motion at each location has been estimated using Shakemap ground motion mapping for each earthquake provided by the US Geological Survey (USGS).

Fragility curves such as those shown in Figure 5.25 can be derived from available post-earthquake damage data for the range of types of building affected, at each level of damage, at a global or a regional level. The curves are drawn using an assumed shape, and finding the curve which best fits each dataset. The authors have been involved in many such studies over a number of years using the damage data assembled in the Cambridge Earthquake Impact Database (CEQID) http://www.ceqid.org.

Such curves can be used for making risk assessments, for example for insurance purposes (Spence et al. 2008), or for estimating the extent of losses immediately after an earthquake has occurred, to assist in planning emergency relief operations. However, there is considerable deviation of the data points from the curves, and this uncertainty needs to be taken into account in using them for financial or operational decision-making.

An alternative approach to vulnerability assessment is to allocate building construction classes to one of a limited number of vulnerability classes, for each of which the relationship between damage and ground motion is pre-defined. This is the approach used in the definition of macroseismic intensity using the EMS-98 scale. In this scale, six vulnerability classes A to F are defined with class A being the most vulnerable, class F the least. The definition of each EMS-98 intensity level partly derives from the expected damage to buildings of different vulnerability classes which will occur at that level of intensity. For example for Intensity VIII (Heavily damaging), according to the EMS-98 definition:

- Many buildings of vulnerability class A suffer damage of grade 4, a few of grade 5
- Many buildings of vulnerability class B suffer damage of grade 3, a few of grade 4
- Many buildings of vulnerability class C suffer damage of grade 2, a few of grade 3
- Many buildings of vulnerability class D suffer damage of grade 2

The term 'few' is defined to mean less than about 15%, and term 'many' to mean 15–55%. Numerical fragility curves in terms of macroseismic intensity can be defined from this set of definitions (Lagomarsino and Giovinazzi 2006), as shown in Figure 5.26. In Figure 5.26, the vertical scale is a composite damage index deriving from the proportion of buildings with damage at each damage level.

Using the data in Figure 5.26, it is possible to estimate the proportion of buildings in each vulnerability class which can be expected to suffer heavy damage or worse at each intensity level, Table 5.3.

Thus, at intensity VI, which is roughly the level of the 500-year earthquake in moderate risk areas, 7% of buildings of class A will suffer heavy damage, but very few buildings of class B and none of classes D and above will be heavily damaged. By contrast at intensity VIII, which is roughly the level of the 500-year earthquake in high-risk areas, the proportions of heavily damaged buildings will be about 70% of class A, 31% of class B, 7% of class C and 1% of class D. Again, buildings of classes E and F will suffer nothing more than moderate damage.

The vulnerability table in Figure 5.27 shows the most likely vulnerability class (A–F) for each of the structural classes of buildings which are shown in Table 5.2 and have been discussed in more detail in the chapter. In this table, reinforced concrete buildings, both framed and shear-wall buildings, are divided according to their level of earthquake-resistant design, and EMS-98 gives further guidance about how these terms should be interpreted.

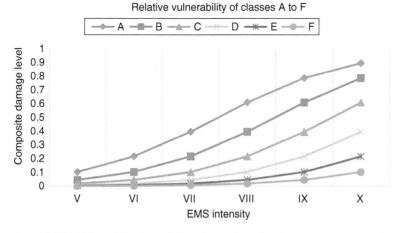

Figure 5.26 Vulnerability for building classes A–F, showing expected composite damage level (max 1.0) at EMS Intensity levels V–X. (The composite damage level derives from the proportions of buildings in each damage survey dataset with damage levels D1–D5). *Source:* Modified from Lagomarsino and Giovinazzi (2006).

Table 5.3 Proportion of buildings of each vulnerability class expected to suffer heavy damage or worse (D3, D4 or D5) at each intensity level.

Vulnerability class	A (%)	B (%)	C (%)	D (%)	E (%)	F (%)
Intensity VI	7	1	0	0	0	0
Intensity VII	31	7	1	0	0	0
Intensity VIII	70	31	7	1	0	0
Intensity IX	93	70	31	7	1	0

The table in Figure 5.27 shows that for any structure type, there is a wide range of possible vulnerability classes. Apart from the level of earthquake-resistant design, which is taken into consideration only for reinforced concrete buildings, the vulnerability of any particular building or group of buildings may be affected by better or worse than average level of workmanship and quality of construction; by the age and state of preservation of the buildings; by the regularity or otherwise of buildings both in plan and elevation; by the position of the building in relation to other buildings; and by whether or not retrofitting has been applied to improve a building's earthquake resistance.

The building classes with the highest expected vulnerability, class A, are the weak masonry buildings, particularly adobe and rubble stone masonry. Other forms of masonry are somewhat less vulnerable, classes B and C, except for reinforced and properly confined masonry which is expected to be of class D. An adequate performance, classes D, E, or F, can be expected from timber structures and from reinforced concrete structures with moderate or better level of earthquake-resistant design.

Defining structural vulnerability is at best, a rather imprecise matter. Structure type is an important indicator of likely vulnerability, but the behaviour of even similar buildings will

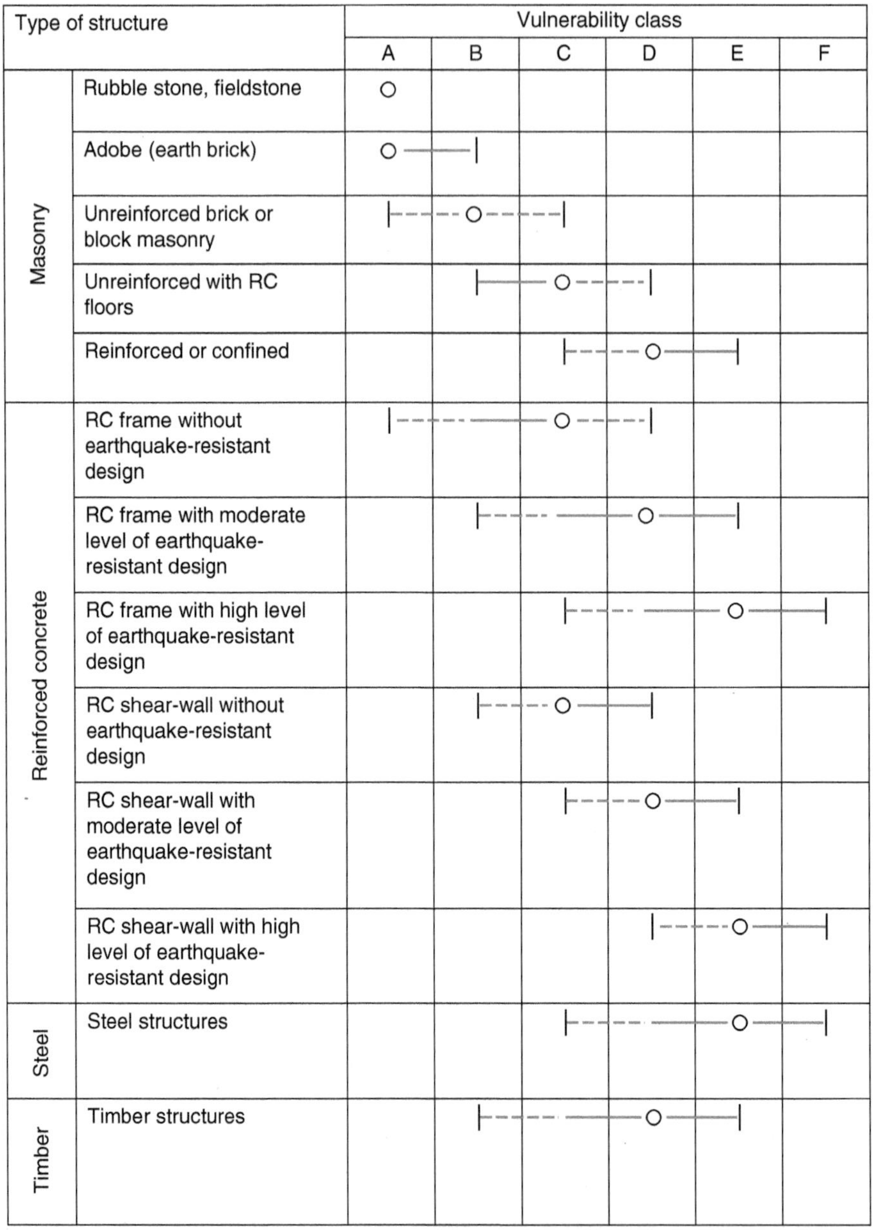

Type of structure		Vulnerability class					
		A	B	C	D	E	F
Masonry	Rubble stone, fieldstone	O					
	Adobe (earth brick)	O—————|					
	Unreinforced brick or block masonry	|——————O———————|					
	Unreinforced with RC floors		|————O——————|				
	Reinforced or confined			|——————O————|			
Reinforced concrete	RC frame without earthquake-resistant design	|————————————O——————|					
	RC frame with moderate level of earthquake-resistant design		|——————————————O————|				
	RC frame with high level of earthquake-resistant design			|——————————————O————|			
	RC shear-wall without earthquake-resistant design		|——————O————|				
	RC shear-wall with moderate level of earthquake-resistant design			|——————O————|			
	RC shear-wall with high level of earthquake-resistant design				|——————————O————|		
Steel	Steel structures			|——————————————O————|			
Timber	Timber structures			|——————————————O————|			

Figure 5.27 Most likely vulnerability range (solid lines) and possible vulnerability range (dotted lines) for the principal classes of building described in Table 5.1. *Source:* Based on EMS-98, Grünthal et al. (1998).

vary apparently randomly, even when subjected to the same shaking level. Matters such as the duration and frequency content of the ground motion, the nature of the subsoil and aspects of the construction quality including the foundations, will all have an effect which is only partly predictable. Nevertheless, we can see clearly from Figure 5.27 just how much more vulnerable the weak masonry buildings (classes A and B) are compared with buildings of reinforced concrete or steel built to modern codes or standards (classes E and F).

Unfortunately, as seen in Chapter 3, it is the more vulnerable building types which are also most numerous in many of the world's most earthquake-prone regions, leading to the risk of future high-casualty disasters. Accordingly, Chapter 6 will look at the human casualty implications of building vulnerability, and Chapter 7 will consider what can be and is being done to improve building safety around the world.

References

Alimoradi, A. (2002). *Steel Frame with Semi-Rigid Khorjini Connections and Jack-Arch Roof*, Report 25. Iran: World Housing Encyclopedia.

Alvarez Velasquez, C., Hube Ginestar, M., and Rivera Jofre, F. (2016). *Confined Masonry*, Report 181. Chile: World Housing Encyclopedia.

Arnold, C. (2002). *Wood Frame Single Family House*, Report 65. USA: World Housing Encyclopedia.

Arnold, C. (2004). Timber construction. In: *WHE Summary Publication* (eds. S. Brzev and M. Greene). California: World Housing Encyclopedia, Earthquake Engineering Research Institute.

Arze, E. (2002). *Steel Frame Buildings with Shear Walls*, Report 3. Chile: World Housing Encyclopedia.

Astroza, M., Moroni, O., Brzev, S., and Tanner, J. (2012). Earthquake Spectra Seismic performance of engineered masonry buildings in the 2010 Maule earthquake. *Special Issue on the 2020 Chile Earthquake* 28: S385–S406.

Blondet, M., Villa Garcia, G.M., Brzev, S., and Rubinos, A.W. (2011). *Earthquake-Resistant Construction of Adobe Buildings: A Tutorial*. World Housing Encyclopedia.

Booth, E. (2014). *Earthquake Design Practice for Buildings*, 2e. London: ICE Publishing, Institution of Civil Engineers.

Bothara, J. and Brzev, S. (2011). *Improving the Seismic Performance of Stone Masonry Buildings*. World Housing Encyclopedia.

Brzev, S. (2004). Precast concrete construction. In: *WHE Summary Publication* (eds. S. Brzev and M. Greene). Oakland, CA: World Housing Encyclopedia, Earthquake Engineering Research Institute.

Brzev, S., Scawthorn, C., Charleson, A.W. et al. (2013). *GEM Building Taxonomy Version 2.0*, Technical Report 2013-02 V1.0.0. Pavia: GEM.

Brzev, S. 2020, personal communication.

Buchanan, A., Carradine, D., Beattie, G., and Morris, H. (2011). Performance of houses during the Christchurch earthquake of 22 February 2011. *Bulletin of the New Zealand Society for Earthquake Engineering* 44: 342–357.

Bureau of Indian Standards (2003). *Improving earthquake resistance of low strength masonry buildings*, IS13828. New Delhi, India: Bureau of Indian Standards.

Cardona, O.-D. (2019). Survey response.

CEN (2004). *Design of structures for earthquake resistance, EN 1998-1, section 9: specific rules for masonry structures*. Brussels: European Committee on Standards, European Commission.

Coburn, A. and Spence, R. (2002). *Earthquake Protection*. Wiley.

Coburn, A., Hughes, R., Pomonis, A., and Spence, R. (1995). *Technical Principles of Building for Safety*. London: Intermediate Technology Publications.

D'Ayala, D. (2004). Unreinforced brick masonry construction. In: *WHE Summary Publication*. Oakland, CA: World Housing Encyclopedia, Earthquake Engineering Research Institute.

EEFIT (1994). *The Northridge California Earthquake of 17 January 1994: A Field Report by EEFIT*. London: Institution of Structural Engineers.

EEFIT (2010a). *The Haiti Earthquake of 12th January 2010: A Field Report by EEFIT*. London: Institution of Structural Engineers.

EEFIT (2010b). *The Mw 8.8 Maule Chile earthquake of 27th February 2010: A Preliminary Field Report by EEFIT*. London: Institution of Structural Engineers.

EEFIT (2011). *The Mw 6.3 Christchurch New Zealand Earthquake of 22nd February 2011: A Field Report by EEFIT*. London: Institution of Structural Engineers.

EEFIT (2016a). *The Mw 6.2 Amatrice, Italy Earthquake of 24 August 2016*. London: Institution of Structural Engineers.

EEFIT (2016b). *The Kumamoto, Japan Earthquakes of 14 and 16 April 2016: A Field Report by EEFIT*. London: Institution of Structural Engineers.

FEMA (2006). *Techniques for the Seismic Rehabilitation of Existing Buildings*, 547. FEMA.

Foulser-Piggott, R. and Spence, R. (2013). Extending EMS-98 for more convenient application outside Europe II: development of the International Macroseismic Scale, paper no. 382. In: *Vienna Congress on Recent Advances in Earthquake Engineering and Structural Dynamics 2013 (VEESD 2013)* (eds. C. Adam, R. Heuer, W. Lenhardt and C. Schranz). Vienna, Austria.

Government of Colombia (2011). Colombian seismic building code NSR-10. Government of Colombia.

Government of Nepal (1994a). *Guidelines for Earthquake-Resistant Construction: Low Strength Masonry: Nepal Building Code NBC 203:1994*. Ministry of Physical Planning and Works, Government of Nepal.

Government of Nepal (1994b). Nepal national building code NBC 109: 1994. Masonry: unreinforced, Minsistry of Physical Planning and Works, Government of Nepal.

Government of New Zealand (2017). Seismic assessment of existing buildings, section c8: unreinforced masonry buildings. http://www.eq-assess.org.nz/assets/2017-07/Overview_of_ the_Seismic_Assessment_Guidelines_20170703.pdf (accessed October 2020).

Grünthal, G., Musson, R., and Schwarz, J. (1998). *The European Macroseismic Scale 1998*. Luxembourg: Council of Europe.

Hackmayer, L.C., Abrahamczyk, L., and Schwarz, J. (2002). *Reinforced masonry building: clay brick masonry in cement mortar*, Report 175. Colombia: World Housing Encyclopedia.

Hamburger, R. (2003). Building code provisions for seismic resistance. In: *Earthquake Engineering Handbook* (eds. W.F. Chen and C. Scawthorn), 11.1–11.28. CRC Press.

Jackson, R. (1960). *Thirty Seconds at Quetta*. Evans Brothers.

Jain, S.K. (2016). Earthquake safety in India: achievements, challenges and opportunities. *Bulletin of Earthquake Engineering* 14: 1337–1436.

Jaiswal, K. and Wald, D.J. (2014). *Creating a Global Inventory for Earthquake Loss Assessment and Risk Management (Created in 2008, Revised 2014)*, Open File Report 2008-1160. Reston, VA: US Geological Survey.

Jong, G.T.T., Kyoung, C.K., and Choi, K. (2017). Seismic performance of reinforced concrete columns retrofitted by various methods. *Engineering Structures* 134: 217–235.

Lagomarsino, S. and Giovinazzi, S. (2006). Macroseismic and mechanical models for the vulnerability and damage assessment of current buildings. *Bulletin of Earthquake Engineering* 4: 415–443. https://doi.org/10.1007/s10518-006-9024-z.

Moroni, O. (2004). Concrete shear-wall construction. In: *WHE Summary Publication* (eds. S. Brzev and M. Greene). California: World Housing Encyclopedia, Earthquake Engineering Research Institute.

Moroni, O. and Gomez, C. (2011). *Report 4: Concrete Shear Wall Buildings*. Chile: World Housing Encyclopedia.

Papanikilaiou, A. and Taucer, F. (2004). *Review of Non-engineered Houses in Latin America with Reference to Building Practices and Self-Construction Projects (No. EN21190 EN)*. Ispra: ELSA, European Commission, Joint Research Centre.

Rael, R. (2009). *Earth Architecture*. Princeton, NY: Princeton University Press.

Ramage, M., Foster, R., Smith, S. et al. (2017). Super tall timber: design research for the next generation of natural structures. *The Journal of Architecture*. 22 (1): 104–122. https://doi.org/10.1080/13602365.2016.1276094.

Rodriguez, M. (2004). Confined masonry construction. In: *WHE Summary Publication*. Oakland, CA: World Housing Encyclopedia, Earthquake Engineering Research Institute.

Sherstobitoff, J., Cajaio, P., and Adebar, P. (2012). Repair of an 18-story shear wall building damaged in the 2010 Chile earthquake. *Earthquake Spectra* 28: 335–348.

Spence, R. (1971). *Making Soil-Cement Blocks*. Lusaka: Commission for Technical Education, Government of Zambia.

Spence, R. (1983). A note on earthquake vulnerability in Quetta. *Disasters* 7: 91–93.

Spence, R. (2010). Archiving and reassessing earthquake damage data. *Presented at the 14th European Conference on Earthquake Engineering* 2010, European Association for Earthquake Engineering, Ohrid.

Spence, R. and Coburn, A. (1987). Earthquake protection: an international task for the 1990s. *The Structural Engineer* 65A: 290–296.

Spence, R. and Cook, D. (1983). *Building Materials in Developing Countries*. Wiley.

Spence, R., So, E., Jenny, S. et al. (2008). The Global Earthquake Vulnerability Estimation System (GEVES): an approach for earthquake risk assessment for insurance applications. *Bulletin of Earthquake Engineering* 6: 463–483. https://doi.org/10.1007/s10518-008-9072-7.

Tomazevic, M. (1999). *Earthquake-Resistant Design of Masonry Buildings*. London: Imperial College Press.

Tuna, Z. and Wallace, J.W. (2019). Collapse assessment of the Alto Rio Building in the 2010 Chile earthquake. *Earthquake Spectra* 31 (3): 1397–1425. https://doi.org/10.1193/060812EQS209M.

Yakut, A. (2004). Reinforced concrete frame construction. In: *WHE Summary Publication*. Oakland, CA: World Housing Encyclopedia, Earthquake Engineering Research Institute.

6

How is the Population Affected?

Earthquakes do not kill people, collapsing buildings do. Building collapses were responsible for over 75% of all casualties in global earthquakes in the period between September 1968 and June 2008 (Marano et al. 2010) and continue to be the dominant causes of deaths in earthquakes (Daniell et al. 2012; So et al. 2018). Although there are many success stories, some highlighted in this book, of how engineered buildings have saved lives against this natural hazard, many preventable deaths from earthquakes are still occurring around the world. Given low-cost and informal buildings are most likely to fail, earthquakes will also continue to disproportionately affect the poorest in the community.

6.1 Causes of Earthquake Casualties

A recent detailed analysis (So et al. 2018) of the causes of deaths and injuries in a human casualty dataset was assembled from 39 non-tsunamigenic earthquakes. The 39 events are from 1906 to 2012 and include 19 earthquakes in Europe, 14 earthquakes in the USA, 2 in Japan, 1 in each of Algeria, Indonesia, New Zealand and India. The magnitudes range from Mw5 to Mw7.8.

The data showed that building collapse is the most common cause of death within the dataset (Figure 6.1), accounting for 78% of the 1193 recorded deaths. The next highest at 6.3% was falling masonry and/or glass, showing the importance of considering non-structural elements in building design in seismogenic regions. This was reiterated by the analysis of the injuries caused inside and outside, which showed that 75% of the recorded injuries were inside a building; 10% were outside (the remaining were either unknown or in a car).

From the same study, the most common causes of minor injuries indoors were either from falling over or being thrown from a chair or a bed and being hit by falling objects (non-structural). Outdoor injuries caused by falling masonry, glass or other debris were also significant. In terms of injuries (for both indoor and outdoor) categorised in the database, 68% were sprains; 17% were cuts and bruises; 7% were broken bones. Other types of injuries less significant in terms of quantity in the database included coma; amputation; burns; and head injuries.

To create this database, data on fatalities and serious injuries was collected through a systematic review of academic papers, insurance claims, newspaper articles, personal accounts

Why Do Buildings Collapse in Earthquakes?: Building for Safety in Seismic Areas,
First Edition. Robin Spence and Emily So.
© 2021 John Wiley & Sons Ltd. Published 2021 by John Wiley & Sons Ltd.

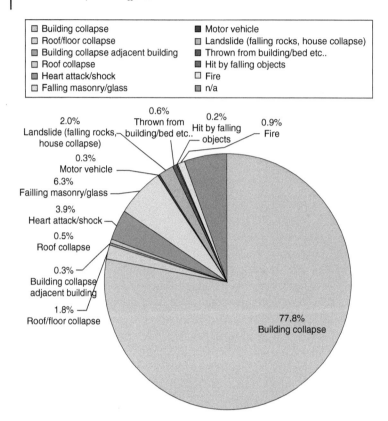

Figure 6.1 Chart showing the causes of deaths from the casualty database. Building collapse accounted for 78% of the 1193 recorded deaths. *Source:* Author's own.

and other open source websites for each event. The study did not include tsunamigenic earthquakes occurring offshore and causing loss of life primarily due to drowning, as the focus was on casualties related to the action of ground shaking on buildings and people.

The dataset contained qualitative descriptions of over 8300 individual deaths and injuries including information about the level, type and place of injury; whether the person was inside or outside, the cause of injury, the type of injury, body parts affected, and whether the injury was caused by the main event, an aftershock or during the clean-up operations. It should be noted that 6877 of the 8300 data points (83%) are from the Canterbury earthquake sequence (Mw7.0 Darfield earthquake in 2010 and Mw6.1 Christchurch earthquake in 2011) which may skew the data. However, the New Zealand earthquakes only contributed to 4% of the deaths shown in the graph in Figure 6.1.

The investigation highlighted the main causes of deaths and injuries from earthquake ground shaking. The fact that building collapses are still the major cause of deaths and injuries comes as no surprise; however, in order to fully understand earthquake casualties, one needs to delve deeper into the influencing factors causing a building's response and that of its occupants. Figure 6.2 illustrates the main factors contributing to the final casualty number of a seismic event. Some are quantifiable-type of ground motions, building responses and number of people inside buildings; others are harder to predict, such as behavioural aspects and the efficiency and accessibility of search and rescue. Well-coordinated and rapid search and

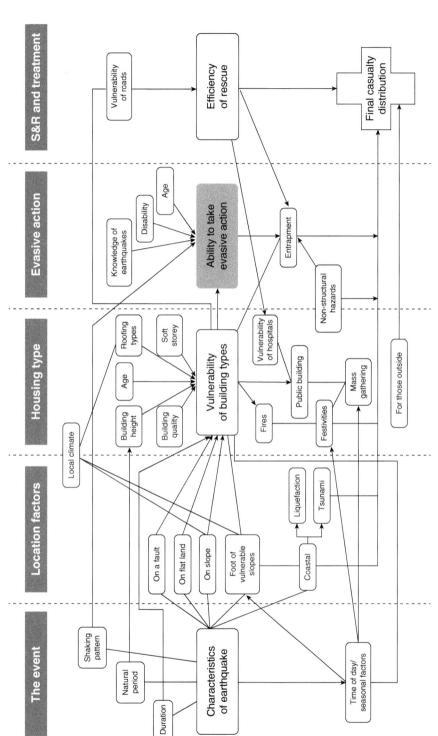

Figure 6.2 Flow-chart showing the different causal factors to consider when investigating and modelling casualties due to earthquakes. *Source*: Author's own.

rescue efforts could be vital to the survival of trapped occupants. These aspects will be taken and discussed in turn in this chapter, and their impact on the final casualty number will be reviewed. An overview of casualty estimation in earthquake loss modelling including a review of the current methods of modelling, the shortfalls of models and reasons behind these shortfalls, as well as some insights from the field will be presented. In doing so, we shall be highlighting some aspects of casualties in earthquakes which are currently not considered in loss models but may be significant to the final death and injury numbers in future events.

6.2   Casualties due to Building Collapses

6.2.1   How are Occupants Affected?

As explained in the earlier chapters, the way a building responds to an earthquake varies depending mainly on the characteristics of the motion, site conditions, the building materials and its structural arrangement. As the seismic wave energy is transferred to the buildings, the motion experienced by the occupants indoors also varies. For example, occupants in a high-rise building such as workers in a Tokyo office tower would feel slow swaying as the building rocks from side to side due to the Mw9.1 Great Tohoku earthquake and tsunami in 2011 some 300 km away, whereas residents in a low-rise unreinforced masonry building in the immediate vicinity of an earthquake epicentre will experience very strong horizontal and possibly vertical movements such as in the case of the Mw7.9 Wenchuan earthquake in 2008.

The way buildings respond to earthquakes is briefly explained in Chapters 4 and 5. In earthquake engineering design, there are strict limits in codes to the amount of movement a building can make. This is mainly to do with occupants' comfort levels. While in Tokyo, the occupants may feel unsettled by the swaying of the buildings as a result of the Mw9.1 Great Tohoku earthquake and tsunami in 2011, strong motions occurring directly under vulnerable housing will kill, like in the Mw7.9 Wenchuan earthquake in 2008.

6.2.2   Can Building Occupants Escape?

The ability of occupants to move during an earthquake depends on the fitness of the resident, the size of the event and distance of the building to the epicentre, the type and severity of motion (swaying versus vertical motion), the building's structural type and the topography on which the building is founded. Evasive actions, whether it is to drop, cover and hold or running outside of a vulnerable structure, have worked well in the past only when the inhabitants can act fast, usually in seconds between the primary and secondary wave phases of the seismic motion (see Chapter 4).

We are often asked what is the best thing to do during an earthquake. Unfortunately, this is not an easy question to answer. The most appropriate evasive action to take will depend on the location and situation of the affected people during the earthquake. The safest action will depend on the level of shaking, building type, a person's location within the building and access to exits. If the building collapses very quickly, as can happen with poor construction, people may not have an option to act at all. GeoHazards International (GHI) published a manual (Figure 6.3) in 2015 which was revised in 2018 (GHI 2018) to provide guidance on what people should do during earthquake shaking to protect themselves from

Implementation Workbook

Developing Messages for Protective Actions
to Take During Earthquake Shaking

March 2018

Figure 6.3 GHI's evidence-based guidance describes the process to use and the key considerations for creating effective messages for different contexts. *Source:* Credit: GHI.

injury or death. This document is a compilation of input from subject experts from many earthquake-prone countries around the world. Its focus is helping practitioners develop the right messages to convey to local people on action to be taken during an earthquake. Given the variability in vulnerability of a person's location and situation, this manual is especially welcome in the field as one size does not fit all in this instance.

As part of a study into causes of injuries and death, we carried out two 500-household questionnaire surveys in Pakistan and Indonesia after the Mw7.6 Kashmir earthquake in 2005 and the Mw6.3 Yogyakarta earthquake in 2006. The low death toll relative to the number of destroyed houses and other buildings in Indonesia that occurred at 05:54 local time was attributed to swift evasive action in a devastating event in terms of building damage. What we found was that people were residing in vulnerable masonry housing with timber roofs, built on flat land and well-spaced from each other. The ability to run outside to flat open ground before the buildings collapsed saved lives. The same actions during the Mw7.6 Kashmir earthquake in 1995, in Balakot in Pakistan would not have been possible, due to the town's proximity to the causative earthquake fault, the presence of a strong vertical motion amplified by the ridge

Figure 6.4 Figure showing collapsed buildings in Balakot, Kashmir, after the 2005 earthquake. Heavy concrete roofs/slabs came off failed load-bearing masonry walls trapping residents or slid down the slope when close to the ridge. The red line shows the active fault. *Source:* Author's own.

effect, and very heavy reinforced concrete roofs that crushed inhabitants instantaneously. There was no time, and the motions were too strong to run outside.

The photograph in Figure 6.4 shows the aftermath of the Kashmir earthquake, with flat concrete slabs sliding to adjacent dwellings on the steep terrain. No evasive actions were possible in this instance, and sadly, in this village alone, over 80% of its residents died.

6.3 Survivability of an Occupant in a Building

It is useful to narrate this section from the viewpoint of a person affected by an earthquake. This was the approach used to devise questionnaires and face-to-face interviews with survivors from three earthquakes (So 2009). The questionnaires captured information from the moment the earthquake struck – where the survivors were, what they were doing at the time to where they were at the time of the interview, several months later. It continues with questions on what happened to their houses and what injuries they sustained at the time of the event as well as afterwards. These rich personal accounts provide insights on the contributing factors to deaths, injuries and survival not captured by other means, and help inform modelling of earthquake casualties, as well as focus on Search and Rescue (SAR) and educational efforts in earthquake preparedness.

In total, over 1200 questionnaire forms were returned for three events, namely the Mw7.6 Kashmir earthquake in 2005, the Mw6.3 Yogyakarta earthquake in 2006 and the Mw8.0 Pisco earthquake in 2007. These were all earthquakes in developing countries, where there was widespread damage due to a combination of vulnerable housing under intense shaking. These three earthquakes are typical in so far as the damage was expected, given the seismic intensity levels experienced, but there were some surprising anomalies which can be investigated and explained with these surveys. For example, in Pakistan, there were notably more serious injuries surviving beyond the expected period. The earthquake affected a mountainous part of north-eastern Pakistan which damaged infrastructure while landslides blocked access to remote towns and villages for days after the event. There is obvious bias as the surveys were carried out on survivors of the event and therefore those that have perished due to serious injuries were not included. However, the survivability of the seriously injured was attributed by local doctors to the natural resilience of the affected population to harsh environments.

By contrast, the death tolls in Indonesia and Peru were disproportionate to the amount of damage observed. There was widespread damage in the affected areas of both Yogyakarta and Pisco earthquakes. The Mw6.3 Yogyakarta earthquake in 2006 affected over 8 million people in this densely populated region of Java, and over 150000 houses were destroyed, and nearly a quarter of homes experienced damage. The earthquake impacted five districts within Yogyakarta province and six within neighbouring Central Java province. A total of 2.7 million people lived in the houses that were destroyed or damaged. 1.5 million were made homeless as a result (OCHA 2006). However, despite the extensive structural damage, less than 6000 people lost their lives. In the surveys that were carried out, the participants attributed their survival to the ability to run outside onto flat land whilst or before the roof of the buildings collapsed.

After the Mw8.0 Pisco earthquake in 2007, over 50000 houses were completely damaged, mainly single-family residential adobe dwellings. A total of 519 lost their lives in this earthquake. At the epicentre in the city of Pisco, 80% of the town's adobe housing collapsed or sustained heavy damage. There were several reasons why there was such widespread damage. First and foremost, the quality of the building stock in this area was poor. Traditional earth structures, those of adobe construction, designed without the inclusion of any kind of reinforcement, and without consideration of earthquake-resistant concepts, were susceptible to collapse in a sudden, brittle way. The adobe blocks and mortar in Pisco and surrounding areas were made with sandy soil, which did not have enough clay to provide good adhesion between mortar and adobe blocks.

Despite this devastation, less than 1% of Pisco's population was killed in residential buildings and 2.5% injured. In all, 430 residents in Pisco died, including 150 that perished when the roof of the San Clemente Church collapsed during a funeral service. Of the 200 households interviewed in the affected area, 30% were in collapsed buildings but had no injuries at all. This contradicts other empirical studies for adobe housing. For example, after Mw6.6 Bam earthquake in 2003, about 39% of the people in collapsed adobe buildings who were not killed, were injured (Kuwata et al. 2005).

This could be attributed to three factors: all the time history recordings in the Pisco area showed a total duration of shaking of approximately 160 seconds with three sequences of motion with a period of 20–30 seconds of smaller amplitude in between the two stronger parts. This calmer interval allowing people indoors to escape from their low-rise houses was one of the reasons given for survival during the interviews. The time of the earthquake (18:40 local time), as well as the extent of collapse of the local buildings related to the light weight and type of roofs could also be reasons for the low death toll. The adobe buildings in Bam had very thick compacted earth roofs.

These atypical facts make modelling casualties with generalised assumptions difficult, and we have still to account for people's behaviour during an earthquake.

When an earthquake strikes, depending on the characteristics of the event, the affected person's understanding of their situation and therefore their behaviour, their survivability in a building will differ. This does not account for the vulnerability of the building itself.

A person's behaviour is difficult to assess and may contradict their own earthquake awareness and common sense. During our reconnaissance missions to Pakistan after the Mw7.6 Kashmir earthquake in 2005 and a recovery mission to Japan after the Mw9.1 Great Tohoku earthquake in 2011, we spoke to survivors to gain an insight of their experiences and what contributed to their survivals. In Pakistan, the earthquake happened at 8:50 in the morning, schools were open, and most people were out of their houses working or on the streets. The people who were indoors, inside vulnerable unreinforced masonry houses were the elderly and children in poorly built schools. According to government figures, 18 095 students and 853 teachers died in the earthquake, most of them in widespread collapses of school buildings. The earthquake disproportionally killed more women and children. Accounts from survivors of women deliberately running into harm's way to try to extract young children from danger could be one explanation. In Japan, the same familial instincts also saw people running towards the coast, against their ingrained knowledge of heading for higher grounds as sirens were warning of an impending tsunami. Among the survivors, who recounted their predicaments and how lucky they were to have survived, some would say how they knew it was the wrong thing to do, but they had to make sure their family members were safe.

These stories often go undocumented but are reminders that the underlying cultural, personal and social contexts are important when considering earthquake risk reduction programmes. Unfortunately, due to the small number of anecdotes, the findings do not form a representative sample to establish a general theory; but collated, these accounts from all over the world allow us to gain a better understanding of the challenges people face during earthquakes.

The personal stories do not negate the overwhelming evidence that earthquake preparedness in the form of better buildings and evasive actions of drop, cover and hold during an earthquake save lives, as witnessed recently in Mw8.8 Maule earthquake in 2010.

Chile has had a dozen earthquakes of magnitude 7.5 or higher since 1970 and has been regenerating its building stock continuously. Unreinforced masonry buildings and soft storey structures have been banned, the former since the late 1930s. Since 1977, emergency managers in Chile have been overseeing *Operacion Deyse*, supporting earthquake drills three times each year in Chilean schools. All schoolchildren, as well as private and public sector employees are taught 'Drop, cover and hold on'. When an earthquake occurs, everyone will drop to the ground, get under a table or door frame, protect their head with their arms and hold on until the shaking stops.

Investigations of a person's behaviour during and after an earthquake are rare. Amongst the first researchers to carry out such surveys were Ohta and Ohashi (1985) and Okada (1996a). They carried out household surveys and interviews of survivors, logging their immediate and subsequent movements. Figure 6.5 below is taken from Okada (1996a) and shows an example of the movements of residents and furniture in their homes during strong shaking. Okada obtained 284 questionnaire responses from the injured in the

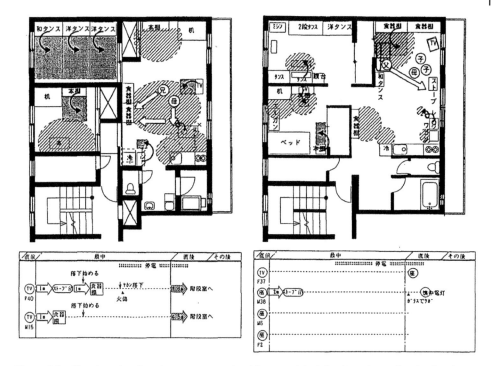

Figure 6.5 Sketches showing the movement of residents and their furniture immediately after the Mw8.3 Kushiro-oki earthquake in 1993. *Source:* Okada 1996a. © 1996, Scientific Information Database.

earthquake-affected area of Kushiro in Japan in 1993. Although the samples collected were not be representative of the whole population, we can learn from them, to assess common behavioural patterns and use the data for spatial planning and earthquake education.

Figure 6.5 shows two sketches of the location and movements (denoted by open circles and arrows) of the occupants in a two-storey house during the Mw8.3 Kushiro-oki earthquake in 1993. The hatched areas show the location of fallen items/furniture after the earthquake. The tables beneath the floor plans show the timeline of what happened to the household after the earthquakes, some noted injuries (during the earthquake/clean up), electricity outages, etc.

More recently, Lambie et al. (2016) explored the potential of using CCTV video footage (outside of buildings) to assess human behaviour from the Mw6.1 Christchurch earthquake in 2011. They argue that research into human behaviour during earthquakes based on interviews and surveys is subject to bias and therefore, a standardised methodology is needed to examine relationships between human behaviour and injuries objectively. They presented the CCTV Earthquake Behaviour coding scheme which uses a set of keywords to capture the sequence of events in the footage, which could tap into another valuable resource for understanding people's behaviour.

The amount of space (volume) available for occupants when trapped but not killed and, of course, the speed and ability of search and rescue teams would have an impact on the survivability of a victim in a collapsed building. It is worth noting that an increased lethality rate due to collapse is likely when the proportion of collapsed structures exceeds a certain threshold, e.g. 30% of total building stock, to account for limitations of search and rescue capabilities (FDMA, 2008, personal communication).

Feng et al. (2014) in their study of estimating earthquake casualties using remote sensing presented a set of graphs showing the change in the survival rate across a duration of 100 hours from the event occurrence of occupants in different types of collapsed buildings. They found that ~40–60% of the trapped people in collapsed buildings died in a very short time. The survival rate varied, depending on the attributes and damage grade of buildings. Using data from three earthquakes, namely the Mw6.6 Bam earthquake in 2003, the Mw7.9 Wenchuan earthquake in 2008 and Mw6.9 Yushu earthquake in 2010 in China, they computed the casualty rates and changes in the survival rates in different collapsed buildings, and some of these are shown in Figure 6.6.

The survival rate of what they call unbound aggregate structure (URM) buildings remained low from the start, while the survival rates of occupants in wood-frame buildings were comparatively higher. As expected, these are to do with the collapse mechanisms, weight of materials and possible void spaces for occupants to survive in.

In general, victims are killed in buildings by (So 2009)

1) crushing or suffocation under collapsed structural and non-structural elements;
2) asphyxiation by the volume of dust generated by the collapse;
3) delay in being extricated and rescued.

But for the survivors in a building, how are they trapped?

6.3.1 How are People Trapped?

The flowchart in Figure 6.7 shows the Cambridge casualty estimation model (Coburn and Spence 2002) to calculate the number of human casualties in collapsed buildings following an earthquake. As shown, the model assumes a certain number of people will be trapped, and some will die instantly, and others will need to be rescued. In order to populate the listed parameters, detailed analyses of what happens to each of the occupants in a collapsed building is required. The ability to populate casualty models with numbers from actual events helps with the general understanding of earthquake casualties, and with validation and calibration of loss estimation models.

There are some earthquakes where the death tolls are governed by a handful of cases of catastrophic collapses. This was the case for the Mw7.3 El Asnam earthquake in 1980 in Algeria, where it is estimated that at least 1000 people died, out of 6500 for the entire event, due to the collapse of a single building (the Cite An Nasr Complex) and hundreds due to the collapse of the Grand Mosque (as the earthquake occurred on a Friday at prayer time). In 1983, 25% of all deaths of the M5.5 Popayan earthquake in Colombia was due to the collapse of the Cathedral (Gueri 1983); similarly, the collapse of four reinforced concrete buildings contributed to 366 deaths out of 1500 in the Mw5.7 El Salvador earthquake in 1986. A more recent example is the Cathedral de San Clemente in Pisco which killed 148 people (there were only two survivors from the roof collapse) after the Mw8.0 Pisco earthquake in 2007. Whilst analyses of single building collapses would skew general loss models, they do provide insights into entrapment and the part non-engineering factors play in determining survival from earthquakes.

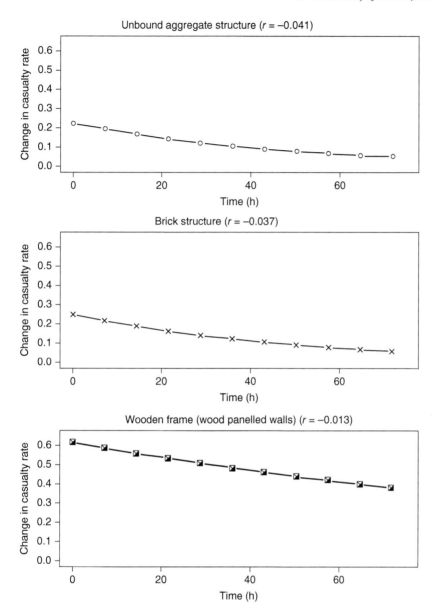

Figure 6.6 Postulated changes in survival rate of occupants in different collapsed buildings.
Source: Feng et al. 2014. Licensed under CC-BY-3.0.

The Mw6.1 Christchurch earthquake in 2011 occurred at 12:51 local time. The collapses of two mid-rise reinforced concrete (RC) buildings accounted for 75% of the final death toll of 181 people. Unlike the Mw7.0 Darfield earthquake in 2010, this event due to its inopportune timing and proximity to the city also caused nearly 6900 injuries as recorded by the Accident Compensation Commission of New Zealand. Figure 6.8 shows the collapse of one of those two buildings, the Pyne Gould Corporation (PGC) building.

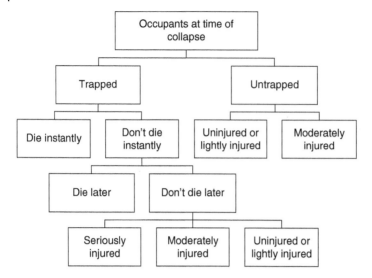

Figure 6.7 Flowchart showing the Cambridge casualty estimation model in collapsed buildings. *Source:* Coburn and Spence 2002. © 2002, John Wiley & Sons.

Figure 6.8 Catastrophic collapse of the five-storey Pyne Gould Corporation (PGC) Building in the Central Business District of Christchurch, New Zealand. *Source:* Flickr.

Table 6.1   Table showing what happened to the occupants inside the two collapsed RC buildings after the Mw6.1 Christchurch earthquake in 2011.

	Occupancy at time of earthquake	Trapped	Untrapped	Dead	Rescued
CTV building	158	137	21	119	39
PGC building	38	37	1	16	22
Combined	196	174	22	135	61

Source: Pomonis et al. 2011, Seismological Society of America.

A detailed analysis of the casualties that occurred in the two collapsed RC buildings including analysis at each floor level, discussion on the injuries suffered by the survivors, duration of search and rescue operations, and survival lessons was conducted by Pomonis et al. (2011). A precise tracking of each occupant of the buildings was possible as there was great interest from the media who reported on the exact locations and time and condition of the extraction of survivors. Table 6.1 is a summary of what happened to the occupants of the two collapsed buildings.

It was found that in the Canterbury Television (CTV) building, 75% of the occupants inside the building at the time of the earthquake were killed as the collapse was catastrophic, and a fire raged in the debris for many hours. One interesting finding made in the study was that there were only 38 people inside the PGC building at the time of the earthquake, but more than 70 could have been inside if the earthquake had not taken place during the lunch break. A similar but less-pronounced effect is valid for the CTV building. It could be argued that perhaps the time of the earthquake was the worst possible for people outdoors in the Central Business District (CBD), but the best possible outcome for those normally inside the buildings during business hours.

From this analysis, we can gain an understanding of not only the survivability of occupants in collapsed, reinforced concrete buildings but also the impact of the time of day of an event. In Christchurch, many of the workers were outside the buildings on their lunch break when the earthquake struck. The death toll may have been higher if the earthquake struck at other times of the workday. On the other hand, the foot traffic in the CBD was higher, and at least 11 people were killed, and many more were injured when the historic unreinforced masonry buildings in the city centre failed, and pedestrians as well as people in two buses were hit by the falling debris.

6.3.2   Falling Building Debris

The importance of considering falling building debris in casualty estimation was reinforced by the Mw6.1 Christchurch earthquake in 2011, where 35 of the 39 people who died as a result of failures of unreinforced masonry were not inside the buildings (Abeling et al. 2018).

There is clear evidence from recent events that the debris from elements such as gable walls, parapets and chimneys can result in casualties to those outside a building. Figure 6.1 indicates that fatalities from these causes are a significant proportion of all fatalities for the earthquakes considered. Though currently subject to wide uncertainties, further work on assessing falling debris, including analytical studies will help improve these proposed relationships (Abeling et al. 2018; So et al. 2018) so that they can be used with confidence in future loss modelling studies.

6.4   Other Causes of Casualties

In Petal's detailed study (2004) of deaths and injuries from the Mw7.6 Kocaeli earthquake in 1999, it was found that 85% of all injuries and deaths in the survey population are attributable to being struck, caught under, cut or pierced by a falling or fallen object. Non-structural causes were therefore significant, though we must account for the fact that the survey was carried out on survivors of the event. The deaths in the survey were reported by the survivors of the same households.

The objects responsible for 70% of the injuries and deaths were in order of frequency: infill walls, ceilings/beams, free-standing cabinets, columns, glass objects, wardrobes, chests of drawers, and broken windows. Splitting the data into structural and non-structural causes, it was found that 22% of injuries were structurally caused, 68% were due to non-structural causes (non-structural building elements and building contents) and 11% due to both. For deaths, 66% of the deaths were from structural causes, 26% non-structural and 13% due to both. These findings corroborate with the Shoaf et al. (1998) surveys of Californian earthquakes, where it was found that 50% of all injuries were caused by non-structural elements and contents. Detailed casualty studies after destructive earthquakes are rather uncommon though research in this field has been growing in the last 30 years. Studies where causes of casualties are split into structural and non-structural are even more scarce with non-structural causes playing a bigger role in areas where building collapse is less prevalent and where the non-structural components are numerous (e.g. in hospitals, schools, offices, restaurants). Japanese, the US and New Zealand investigations of earthquake casualties are probably the best datasets there are in the field due to the impressive rate of returns of their surveys and the frequency of events in the countries.

Outside of the built environment, secondary earthquake hazards such as tsunamis and landslides have dominated headlines in the past two decades.

6.4.1   Tsunamis

Over 18 000 people died in the Mw9.1 Great Tohoku earthquake in 2011 due to the tsunami plus an additional 217 due to building collapse, landslides and the nuclear accident, and more than 222 000 people died in the Mw9.1 Indian Ocean earthquake and tsunami in 2011. With an increasing, urbanising global population and growing urban areas in coastal zones, many more people are at risk of tsunami and other coastal hazards around the world. The death toll in Japan could have been much worse, if people were not prepared and evacuation procedures and structures at higher ground were not in place. Analyses of the types of forces that are exerted on buildings by tsunami waves are rare, but the Tohoku event offered a group of researchers the opportunity to examine the possible failure mechanisms through observations in detail (Macabuag et al. 2018). This knowledge is invaluable for designs in the future, for checking structural capacity and recognising that failures may occur due to a combination of these loads, effects and mechanisms. For tsunamis, the hope is that there is enough time between the earthquake and tsunami arrival, and that people are prepared and able to head to higher ground.

What was alarming about this event was that the painstaking preparedness measures in Japan were not enough to completely mitigate the tragedy. Historic and scientific data used to plan and design the evacuation centres (their locations and heights) could not have anticipated waves exceeding 10 m and reaching up to 30 m that engulfed the east coast of Northern Honshu, showing the limits of loss modelling.

6.4.2 Landslides

Earthquake-triggered landslides can dramatically change the natural landscape and are also deadly to anything in their paths. During the Mw7.6 Kashmir earthquake in 2005 and Mw7.9 Wenchuan earthquake in 2008, many people were buried or crushed by landslides. In these events, an estimate of around 30% of those who died were attributed to landslides, suggesting about 22 000 and 26 500 fatalities, respectively. Unfortunately, studies that link fatalities directly to landslides induced by earthquakes are rare (Keefer 1984, 2002; Nowicki Jessee et al. 2020).

Keefer states that most of these casualties would have been tallied into the overall death toll, and it is, therefore, difficult to distinguish how many are related to this secondary hazard. One of the very few studies available detailing the fatalities associated to individual landslides is from the Wenchuan event. Table 6.2 shows a list of slope failures, their locations and resulting fatalities caused by the Mw7.9 Wenchuan earthquake in 2008 (Yin et al. 2009). In all, 20 slope failures were recorded, and a total of 4962 fatalities were associated with these failures which ranged from slips that killed 30 in Tuanshan-cun Village in Pengzou City to 1600 in the old town of Beichuan.

Table 6.2 An excerpt of slope failure consequence data triggered by the Mw7.9 Wenchuan earthquake in 2008.

Name of the geohazard	Type of geohazard	Location of the geohazard	Volume ($\times 10^4$ m^3)	Fatalities
Chengxi landslide	Slide	Wangjiayan, Old area of Beichuan County Town	480	1600
Yingtaogou landslide	Slide	Chayuanliang-cun Village, Chenjiaba Town, Beichuan County	188	906
Xinbei Middle School landslide	Slide	New area of New County Junior High School, Beichuan County	240	500
Jingjiashan rockfall	Rockfall	Main road of Southern Beichuan County Town	50	60
Xiejiadianzi landslide	Slide	Team 7, Jiufeng-cun Village, Pengzou City	400	100
Liangaiping landslide	Slide	Tuanshan-cun Village, Pengzou City	40	30

Source: Modified from Yin et al. (2009).

Survival within mass land and mud slides caused by earthquakes is uncommon as there are simply no air pockets. Liquefaction and fire-following are two other main causes of damage, deaths and injuries after earthquakes.

6.4.3   Liquefaction

In Daniell et al.'s (2012) review of causes of deaths from earthquakes, liquefaction is said to account for 0.10% of the global fatalities between 1900 and 2012, amounting to around 2500 deaths. Though when compared with other causes, liquefaction has not contributed to many earthquake-related deaths, a recent event has highlighted how, coupled with other hazards, it can also be very deadly. In the Mw7.5 Sulawesi earthquake and tsunami in 2018, a leaking irrigation canal triggered the Petobo and Jono Oge liquefaction which turned into a mudslide, while the fault rupture triggered the Balaroa mudslide. These mudslides killed more than 3000 people (Rossetto et al. 2019). The tsunami killed a further 1000 people, while building collapses accounted for approximately 300 deaths in this event.

6.4.4   Fire Following

Significant casualties have been caused by fires in the past century especially after the Mw8.1 Great Kanto earthquake in 1923 and the Mw6.9 Kobe earthquake in 1995. About 4.7% of 6122 earthquake deaths in Japan in the period 1964–2005 are attributed to fire following including 275 deaths in the Kobe earthquake, where at least 504 bodies were retrieved from the rubble of burned wooden houses though most of these victims were probably crushed or suffocated before they were burnt (Pomonis 2005). Since then there have been upgrades to flammable wooden dwellings, updates to the seismic building codes and reconfigurations of gas distribution networks in these cities and globally to mitigate this hazard.

The contributory factors highlighted in the sections above should be addressed in loss modelling in order to reduce future fatalities. The remainder of this chapter will look at casualty estimation models for earthquakes, and how field data is incorporated in models.

6.5   How Can We Estimate the Number of Injured and Killed in an Earthquake?

Estimation of the casualties caused by earthquakes is a vital element of loss estimation, contributing importantly to planning of post-earthquake relief, mitigation and insurance. Studies examining the causes of casualties started in earnest in the 1980s with Kobayashi in Japan (1981). This was followed by work of scholars at the Hokkaido University including Ohta, Murakami, Sakai, Pomonis and Coburn in quantifying casualties in the late 1980s and a seminal workshop at the John Hopkins University in the USA in 1989. The workshop entitled 'International Workshop on Earthquake Injury Epidemiology for Mitigation and Response' organised by Dr Nick Jones, Dr Eric Noji and others was important for this area of research internationally with most researchers in the field taking part (Jones et al. 1989).

The extensive building damage that occurred in the Mw6.9 Kobe earthquake in 1995 prompted another wave of research into collapse mechanisms of buildings and impacts on its inhabitants by researchers (Murakami 1996; Okada 1996b). In 1997, the Federal Emergency Management Agency (FEMA) in the United States released the first version of the Hazards US (Hazus) standardised tool that uses a uniform engineering-based approach to measure damages, casualties and economic losses from earthquakes in the United States. The casualty data was taken from US earthquakes, including the Mw6.9 Loma Prieta earthquake in 1989 and Mw6.7 Northridge earthquake in 1994. Hazus-MH 3.0 (FEMA 2015) was released in 2015 and is an event tree model with a nationally applicable standardised methodology that now contains models for estimating potential losses from earthquakes as well as floods, hurricanes and tsunamis in the United States. It is widely used to estimate fatality distributions caused by damaged buildings in the United States and worldwide, where there is a lack of data and country-specific models.

The loss estimates, whether presented in the form of numbers of buildings damaged, population affected, injured and killed or as economic costs, help emergency managers gauge the likely impact of earthquakes and prioritise actions and resources. Given the infrequent nature of earthquakes as compared to other natural hazards, there is time for building upgrade programmes. However, the lack of urgency as compared to other perils also makes it difficult to obtain adequate funding. Loss estimates help quantify the issues. In the immediate hours after a real earthquake event, loss estimation models also help national and international emergency agencies plan for relief efforts.

One such model is the US Geological Survey's PAGER (Prompt Assessment for Global Earthquake Response) model. The PAGER system sends out alerts to the international emergency management and aid communities 20 minutes after an event. This system is robust and has been in operation, providing much needed information on the location, likely exposure and impacts of earthquake worldwide since 2010.

The reason for its success is partly to do with the fact that the intended audience do not need precision in numbers but are after a traffic light system to mobilise resources, therefore, the estimates are given in a logarithmic scale. The PAGER alerting scheme uses four alert levels based on both fatalities and economic loss (Jaiswal and Wald 2010a): green (little or no impact); yellow (regional impact and response); orange (national-scale impact and response); and red (international response). Corresponding fatality thresholds for yellow, orange, and red alerts are 1, 100, and 1000, respectively. For damage impact, yellow, orange, and red thresholds are estimated costs reaching US$1mn, US$100mn and US$1bn, respectively. The summary alert level is the higher of the two alert levels. Accompanying text helps clarify the nature of the alert based on experience from past earthquakes in the region, provides context on the total economic losses in terms of the GDP fraction of the country affected and adds regionally specific qualitative information concerning the potential for secondary hazards, such as earthquake-induced landslides, liquefaction, fire following and tsunami.

The simplicity of the way this information is conveyed contrasts with the subtleties of the underlying models used in PAGER. Near real-time estimation of the alert level for a given event depends on many variables, mainly related with the earthquake source, ground motion propagation from the source, exposure and vulnerability models. Each of these components carry uncertainties, which result in a significant final uncertainty of the model estimation, portrayed through histograms of the economic and fatality-based alert-level probabilities (Figure 6.9).

≋USGS
science for a changing world

Earthquake ● **Red**
Shaking **Alert**

GSN **◉USAID**
FROM THE AMERICAN PEOPLE
ANSS

M 7.8, NEPAL
Origin Time: Sat 2015-04-25 06:11:25 UTC (11:56:25 local)
Location: 28.23°N 84.73°E Depth: 8 km

PAGER
Version 9
Created: 9 weeks, 5 days after earthquake

Estimated Fatalities

Red alert for shaking-related fatalities and economic losses. High casualties and extensive damage are probable and the disaster is likely widespread. Past red alerts have required a national or international response.

Estimated economic losses are 8-40% GDP of Nepal.

Estimated Economic Losses

Estimated Population Exposed to Earthquake Shaking

ESTIMATED POPULATION EXPOSURE (k = x1000)		- -*	- -*	10,721k*	84,253k*	40,899k	3,556k	2,885k	12k	0
ESTIMATED MODIFIED MERCALLI INTENSITY		I	II-III	IV	V	VI	VII	VIII	IX	X+
PERCEIVED SHAKING		Not felt	Weak	Light	Moderate	Strong	Very Strong	Severe	Violent	Extreme
POTENTIAL DAMAGE	Resistant Structures	none	none	none	V. Light	Light	Moderate	Moderate/Heavy	Heavy	V. Heavy
	Vulnerable Structures	none	none	none	Light	Moderate	Moderate/Heavy	Heavy	V. Heavy	V. Heavy

*Estimated exposure only includes population within the map area.

Population Exposure

population per ~1 sq. km from Landscan

| 0 | 5 | 50 | 100 | 500 | 1000 | 5000 | 10000 |

Structures:
Overall, the population in this region resides in structures that are highly vulnerable to earthquake shaking, though some resistant structures exist. The predominant vulnerable building types are unreinforced brick masonry and rubble/field stone masonry construction.

Historical Earthquakes (with MMI levels):

Date (UTC)	Dist. (km)	Mag.	Max MMI(#)	Shaking Deaths
1980-07-29	364	5.5	VII(18k)	0
1980-07-29	388	6.5	IX(11k)	100
1988-08-20	244	6.8	VIII(12k)	1k

Recent earthquakes in this area have caused secondary hazards such as landslides and liquefaction that might have contributed to losses.

Selected City Exposure
from GeoNames.org

MMI	City	Population
VIII	Kathmandu	1,442k
VIII	Patan	183k
VIII	Kirtipur	45k
VIII	Bhaktapur	< 1k
VII	Lamjung	< 1k
VII	Khudi	< 1k
VI	Pokhara	200k
V	Muzaffarpur	333k
V	Gorakhpur	674k
V	Patna	1,600k
IV	Dhankuta	22k

PAGER content is automatically generated, and only considers losses due to structural damage. Limitations of input data, shaking estimates, and loss models may add uncertainty.
http://earthquake.usgs.gov/pager

bold cities appear on map (k = x1000)
Event ID: us20002926

Figure 6.9 Summary of the information provided by PAGER in its fifth released version of the OnePAGER (released four hours after the event) for the Mw7.8 Nepal earthquake in 2015.
Source: 2015-04-25 Nepal M 7.8, USGS.

Additional factors and more detailed analysis in the hours and days after the event may change the initial estimations, resulting in subsequent versions of the information released. These changes usually come from

- seismological updates on the earthquake source (location, depth, magnitude and finite fault model) as new records arrive and computations are refined with time;
- the arrival of in situ information from the epicentral area (macroseismic or strong-motion data) that helps fitting the initial model estimations to ground truth;
- adjustments to the initial estimates made by the PAGER team members, based on expert judgement upon review of the specific conditions in the area or to the nature of the earthquake.

Since its public release in September 2010 (up to March 2019), 5136 public alerts have been issued, including 233 yellow or higher alerts. The PAGER alerts have 460 subscribers, which include internal distribution lists of organisations such as FEMA.

The occurrence or not of casualties is dependent on several factors. The variables across the hazard, exposure and vulnerability interactions make it notoriously difficult to estimate casualty figures reliably (Coburn and Spence 2002), not to mention the lack of collected empirical data and standard methodology for collecting and reporting the number of people killed during earthquakes (Spence and So 2013).

There are different ways to model casualties from earthquakes. Here we have categorised them as statistical and engineering-based approaches. Statistic approaches, such as that of PAGER, perform a regression on country-specific data to estimate fatality rates and, like other statistic models, mathematical relationships of ground motion parameters and the resulting casualties from past earthquakes are used to predict future casualties. Engineering approaches are informed by the structural performance of buildings and locations of the population at the time of the earthquake.

6.5.1 Statistical Approaches to Earthquake Loss Modelling

Fundamentally, all statistical models used in earthquake loss estimation seek to relate a set of independent variables to dependent variables (outputs), relying on empirical data to create the model and for validation. For example, a multi-linear regression analysis proposed by Urrutia et al. (2014) uses matrices to physically quantify the direct damages from earthquakes to human beings and structures given a set of independent variables, e.g. the duration, magnitude, and intensity of earthquakes. This is simply illustrated in Figure 6.10.

Other studies have investigated the loss of life patterns in earthquakes worldwide by correlating the number of casualties with magnitude, distance from epicentre and population density through a quantitative regression (Nichols and Beavers, 2003; Samardjieva and Badal 2002; Shiono 1995).

The underlying working PAGER model is an empirical method that takes the historic earthquakes of certain sizes, in certain countries and determines a relationship between these factors and the number of deaths recorded. The central component of this approach is EXPO-CAT (Allen et al. 2009), a catalogue of human exposure to discrete levels of shaking intensity, obtained by correlating an atlas of ShakeMaps with a global population database. The ability to assign a population to a historic earthquake shaking intensity zone by

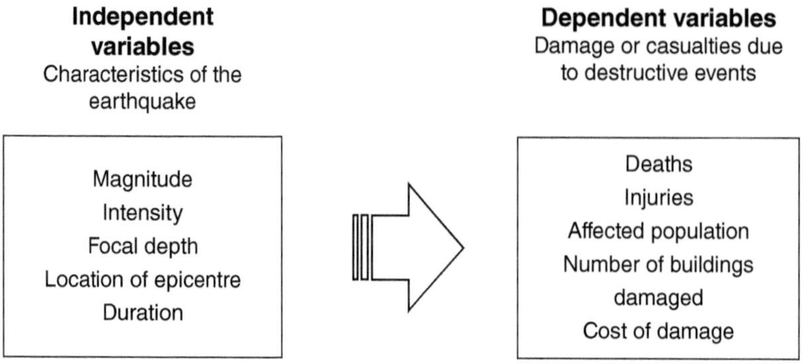

Independent variables
Characteristics of the earthquake

Magnitude
Intensity
Focal depth
Location of epicentre
Duration

Dependent variables
Damage or casualties due to destructive events

Deaths
Injuries
Affected population
Number of buildings damaged
Cost of damage

Figure 6.10 Basic paradigm of linear regression models for earthquake loss and human casualty modelling.

hindcasting and then deriving a fatality relationship for all countries around the world was the ground-breaking work that made PAGER possible. There are however questions on how a past event's death toll can be split reliably by intensity zone and large associated uncertainties arising from this method of assignment.

In addition, the population and the built environment are dramatically changing. An empirically derived model can only be representative of the consequences of the past, based on older building types. For some countries, the PAGER system also relies on a regional model derived using deadly earthquakes from neighbouring countries that have similar building vulnerability profiles in order to infer likely lethality rates. For many countries with limited observations in terms of fatal earthquakes, such assignments can be highly useful, nonetheless, they are subjective and may not be entirely representative of the relative vulnerability of individual country/region (e.g. for Haiti). Another major drawback of purely empirical models is that they could underestimate fatalities in countries where earthquake fatality data are scarce and not recent. Although the underlying empirical model is updated all the time, the PAGER fatality to intensity rates are not time-dependent, i.e. the changes in building stock and their vulnerabilities as well as social vulnerabilities of the current populations are not accounted for. The infrastructure inventory and its vulnerability will change over time as building codes improve, some infrastructures deteriorate and informal housing becomes widespread, for example (Lallemant et al. 2017).

As noted in a recent study on Lima, Peru (Ceferino et al. 2018), PAGER's fatality rates for Peru are estimated from 33 previous earthquakes, in which all the fatalities were under 1000 and only two out of the 33 earthquakes had fatalities exceeding 100. In contrast, data from Japan and Turkey include earthquakes with a larger number of fatalities in recent years; therefore, their estimated fatality rates are higher. The model can only perform a regression on existing data; as large earthquakes are rare events, especially in locations where no large earthquakes have occurred in recent decades (i.e. locations with long seismic gap), there are simply no data points and the postulated fatality rates assumed for those regions will be low.

Other notable statistical approaches include artificial neural network-based (ANN) earthquake casualty estimation models which have been developed recently for specific countries and cities. Aghamohammadi et al. (2013) believe that ANNs can solve and analyse complicated relations stemming from complex seismotectonic properties, building types, social structures, etc., and therefore this is an appropriate method to estimate earthquake casualties in the future.

Daniell and Wenzel (2014) used the analysis of macrosesismc intensity bounds vs. the exposed population and economic exposure to derive empirical fatality functions. An intensive review of over 30 earthquake loss estimation packages and more than 200 casualty estimation methods was carried out by Daniell and Wenzel to generate a set of empirical relationships of intensity to fatality rates which have been calibrated by the socio-economic status of each individual earthquake over the past 50 years. They believe that this is a significant improvement on existing empirical rapid methodologies for earthquake loss studies.

6.5.2 Engineering-Based Approaches

To supplement the broad country-level empirical approach, the PAGER team also runs two engineering-based loss estimation models (a semi-empirical and an analytical model) concurrently with the empirical-statistical model after every event. By utilising knowledge acquired from past events and accounting for the current exposure and vulnerability of the built environment, engineering casualty model estimates help to capture fundamental proxies that tend to dominate the final fatality number. The estimates of fatalities make use of local building inventory, the vulnerability (defined in terms of macro-seismic intensity-based vulnerability relationship to ground shaking and expected damage grades) and consequent collapses of various building types, and fatalities associated with structural collapses of these types (Jaiswal and Wald 2010b).

Other real-time estimation systems include the World Agency of Planetary Monitoring and Earthquake Risk Reduction system (WAPMERR; Wyss 2004) and the European Union Global Disaster Alert and Coordination System (GDACS; de Groeve et al. 2006). These two systems both rely on engineering-based casualty loss estimation approach and therefore require quantification of several important factors, including a database of global building inventory. Figure 6.11 shows the process of a typical engineering-based earthquake loss estimation model.

According to the PAGER team of developers, the empirical model has been the most stable out of the three approaches, perhaps given the uncertainty and number of assumptions made and variables introduced in the other two models, where the reliance on local information and associated uncertainties is far greater. However, given the rapid changes to the built and social environments in earthquake-prone countries, confidence in purely empirical models will diminish and efforts should be focused on improving engineering-based approaches to earthquake loss estimation.

How the affected population are distributed at the time of an earthquake is the first key component of modelling likely casualties using the semi-empirical and analytical approaches. This depends on having an accurate assessment of the number of permanent and temporary residents (commuters and visitors) in the earthquake-affected area at different times of the day. During working hours on a weekday, for example most of the population in an urban setting will be away from their homes and congregated in offices, commercial buildings and schools. Although in most developed countries, there are ample

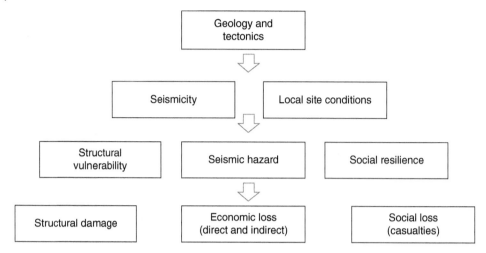

Figure 6.11 Engineering-based earthquake loss estimation approach.

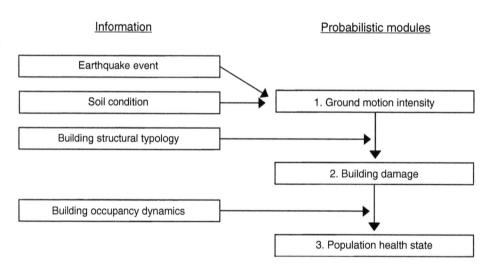

Figure 6.12 Flow diagram showing a multi-severity casualty model framework proposed by Ceferino et al. (2018). *Source:* Ceferino et al. (2018).

data from population and housing census on residential structures and its inhabitants, non-residential building exposures and inventories are hard to develop due to data scarcity. There will also be a proportion of people that is not included in census data including visitors and commuting workers from outside the affected area.

The time of the day and an understanding of the daily routines of the local communities (this would differ depending on culture, season and urban vs. rural environment) would help place the individuals inside or outside of different building types. With this in place, as shown in the flow chart in Figure 6.12 (taken from Ceferino et al. 2018), the vulnerabilities of the different building types are assessed and a fatality to occupancy ratio can be assigned to each damage grade and building type.

In their paper, Ceferino et al. (2018) proposed a regional probabilistic modelling of (i) ground motion intensities correlated over space and across different fundamental periods, (ii) expected damage grades of buildings according to the current building inventory of the region, and (iii) multi-severity casualty occurrence in damaged buildings. Represented in probability modules with associated distributions across the region, each of these components can be updated. This is one of the most promising models to date, though only tested with data from Lima, Peru.

One of the issues with using engineering-based models is that the true risk of collapse of a structure is extremely difficult to estimate through computer analyses, despite the advancement of modelling techniques. Structures are deemed as failed when there is excessive permanent deformation, when brittle failure of certain critical structural components has occurred, or where there is loss of stability to part of the structure. These failure modes do not necessarily result in collapses of structures and furthermore, may not result in casualties.

As previously discussed, even at the same damage grade, the survival rates of occupants inside different building types can vary greatly. This is because the collapse mechanisms of buildings built with different materials and structural form will vary. For example, a timber-frame building, with its light weight and possible voids developing upon collapse will be less deadly than a heavy roofed mud-wall structure. Furthermore, even with the same building material, but varying climatic conditions and construction methods, the potential for causing deaths and injuries will also change. Describing building types only by wall construction material is therefore not sufficient. Even within the same country, the way buildings are constructed will depend on local traditions, most likely affected by the local climate and availability of materials. It was noted in our survey that after Mw8.0 Pisco earthquake in 2007, many people in collapsed adobe houses in this desert area of Peru survived due to the lightweight bamboo mat used in the roofs.

Reinoso et al. (2017) illustrate this point further with the images shown in Figure 6.13. Even in the same building, the location of the occupants would determine their chances of survival as collapses are not uniform.

One of the most comprehensive studies of entrapment and resulting casualties in damaged buildings is from the Mw6.4 Tainan earthquake in 2016. Nine connected 16-storey reinforced concrete buildings known as the Weiguan Jinlong (WJ) Complex collapsed in the city of Tainan in Taiwan. The buildings also fell across a four-lane boulevard and damaged buildings on the other side of the thoroughfare, as shown in Figure 6.14. Pan et al. (2019) examined the injury patterns and entrapped locations of residents inside the WJ Complex, which accounted for 98% of the 115 deaths and 24% of the 513 injuries recorded in the medical incident registry for the entire event. Figure 6.15 has been taken from Pan et al. (2019) and shows the variation in severity of injuries of occupants, on different floors and locations within the complex. Given the variation in the lethality potential buildings, the fatality rates used with the semi-empirical and analytical approaches are possibly the most difficult parameters to quantify. Efforts to estimate fatality distributions in collapsed buildings are presented in Section 6.6.

Figure 6.13 Collapsed buildings where a few may have managed to escape or get rescued immediately (circles), but others died since no space was left (squares) for the emergency units to reach them on time. The arrows point to people working at the top and side of buildings looking for survivors at the Juarez Hospital in Mexico. *Source:* Reinoso et al. 2017.

Figure 6.14 Photograph showing the nine collapsed buildings of the WJ Complex after the Mw6.4 Tainan earthquake in 2016. *Source:* SCMP.

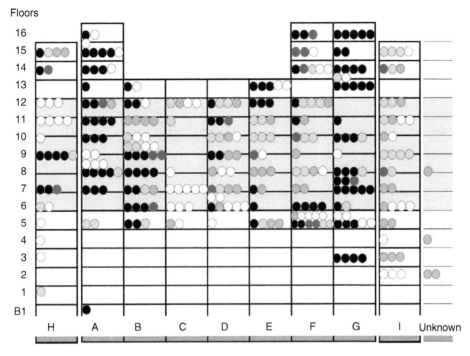

Figure 6.15 Diagram showing the physical locations of patients registered in the Tainan incident registry in the collapsed buildings of the WJ Complex. Black denotes death; red, severe injury; yellow, moderate injury; green, mild injury. White denotes occupants who had not sought medical help within one week of the earthquake. *Source:* Pan et al. 2019.

6.6 Estimating Fatalities Due to Building Collapses

Methods currently used for estimating earthquake casualties tend to be based on assumed lethality rates for buildings in a state of collapse (Jaiswal et al. 2011). However, the collapse damage grade can involve a great variety of different modes of collapse, some involving no more than 10% volume loss and others, such as pancake collapse of reinforced concrete frame buildings, practically 100% volume loss (Jaiswal et al. 2011; Okada and Takai 1999). Lethality rates for the different types of collapse will consequently be widely different (Feng et al. 2014; Pomonis et al. 2011; Seligson et al. 2006; Spence and So, 2013).

The mortality rate of a person inside or outside a building also depends on the quality of the construction, the proximity of a person outside to potential falling debris and the availability of search and rescue and medical care.

6.6.1 Definition of Collapse

The principle of engineering-based methods is that most fatalities are caused by building collapses induced by earthquake ground shaking. The definition of collapse is therefore crucial. Assessing damage to a building and what constitutes a collapse is subjective and has been the topic of debate amongst earthquake engineers for many decades. Using data

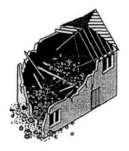

Damage level: DS collapse
Extent of collapse: 10% of volume
M3 (occupants trapped): 10%

Damage level: D5 collapse
Extent of collapse: 50% of volume
M3 (occupants trapped) : 37.5%
(50% of ground floor occupants
assumed escaped before collapse)

Damage level: D5 collapse
Extent of collapse: 100% of volume
M3 (occupants trapped) : 75%
(50% of ground floor occupants
assumed escaped before collapse)

Figure 6.16 Sketches showing the differences in indoor volumetric reduction of a single collapsed building with implications on survivability of its occupants. *Source:* Coburn et al. (1992). © 1992, Taylor & Francis.

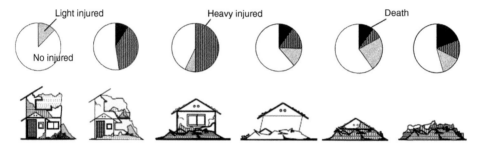

Figure 6.17 Casualty ratios for different collapse patterns of typical Japanese two-storeyed wooden houses (damage grade D5) from Okada and Takai (2000). *Source:* Okada and Takai (1999). © 1999, Architectural Institute of Japan.

collected with such loose definitions of collapse does pose problems. If the definition of complete collapse (D5) of 'more than one wall collapsed or more than half of a roof dislodged or failure of structure members to allow fall of roof or slab' was used (Coburn et al. 1992), the actual volume reduction and therefore lethality potential would vary dramatically. For example, for weak masonry, 'collapsed' buildings can have volumetric reduction ranges from 10 to 100% as shown in Figure 6.16.

The masonry collapses illustrated in Figure 6.16 have an average indoor volumetric reduction of 41%. The survivability of occupants in buildings depends on its collapse mechanism and the volume loss to the structure, as investigated in the study by Okada and Takai (2000). In their study, focus was on damage patterns and a relationship between the building damage patterns and the casualties in each pattern was obtained. The data was based on the field survey carried out in Hokudan-cho Town on Awaji Island which was severely damaged in the Mw6.9 Kobe earthquake in 1995. Figure 6.17 which has been taken from this paper shows that complete destruction of wooden houses was the riskiest of all the collapse patterns in terms of injuries and death distribution among the occupants.

Though of tremendous value, the data in Okada and Takai (2000) was collected from only one event and in order to understand the lethality potential of different buildings, more needs to be done globally.

Given the variation and its implications on casualty numbers and search and rescue (SAR) requirements, an assessment of possible collapse forms of buildings is necessary and was an important component of the work carried out by So (2016). Based on an evaluation of possible collapse mechanisms for load-bearing masonry and framed structures and a careful examination of the least amount of volume reduction that causes fatalities, the definition used for the assignments of fatality rates in Table 6.3 from a collapsed building is as follows:

> At least 10% indoor volume reduction (IVR) at any floor level from whatever cause or mechanism of failure.

The collapse mechanisms would depend on building typologies and the characteristics of the ground motion and the associated lethality within these damaged buildings would depend on the IVR associated with each collapse mechanism. The sketches in Figure 6.18 show the different failure modes of reinforced concrete buildings.

Estimation of the lethality rate based on IVR for different classes of building would clearly be a step forward, but until now systematic collection of post-earthquake collapse and associated casualty data has been lacking. So (2016) developed a set of estimated relationships between lethality rates and IVR for different common building types, from judgement based on reported casualty rates worldwide in areas with different building types being predominant at the time of the examined events, and on experience of the causative collapse mechanisms and the associated IVR (Table 6.3).

Estimates of average fatality rates as a function of average IVR in Table 6.3 have been plotted (Figure 6.19) for timber and masonry buildings. The number of individual buildings for which an estimate of both IVR and the lethality rate among occupants can be obtained is relatively small, but it clearly shows an increase in lethality with IVR; and there are notable differences in the relationship between IVR and lethality for the two highlighted building construction types.

Whilst these figures are preliminary and based to some extent on expert judgement, they suggest that understanding the likely IVR in collapse is crucial to making reasonable estimates of fatality rates, building by building and building class by building class.

Given a lack of studies directly linking building damage and specifically IVR to void space to casualties, So et al. (2018) assembled a damage and casualty dataset to further examine the possible relationship between IVR and casualties for different building types. In the study, a photographic database was compiled using 435 photographs from 47 separate earthquake events utilising a range of sources including the Earthquake Engineering Field Investigation Team (EEFIT) field missions and Earthquake Engineering Research Institute (EERI) datasets.

The photographs in Figure 6.20 show two images from this photographic evidence database. This illustrates clearly the need for examining IVR in casualty studies. Though both structures suffered damage from Turkish earthquakes and were categorised as Damage Grade D5 according to EMS, the lethality potential in these buildings is very different, given the upper stories in the right-hand photograph were relatively intact.

An IVR approach would therefore improve the estimation of fatalities and serious injuries inside buildings. This was the method adopted by Crowley et al. (2017) in examining the risk from induced seismicity to buildings in The Netherlands. In their study, rather than relying on empirical observations, the extent of collapse of the buildings was assessed

Table 6.3 Sample of judgement-based fatality rates for 15 different building types for use in loss estimation models.

	Building classes	Typical IVR (%)	Fatality ranges from literature (%)
	Light timber		
1	With light roof	<10	0.25–1
2	With heavy roof	>50	0.75–3
	Heavy timber		
3	With light roof	<10	0.5–1
4	With heavy roof	>50	2–3
	Weak masonry		
5	Adobe light roof	<50	5–15
6	Adobe heavy roof	>75	20–90
7	Irregular stone with wooden pitched roofs (low-rise)	40–60	5–20
8	Irregular stone low-rise concrete slab roofs	>70	10–40
	Load-bearing masonry		
9	European	>30	3–12
10	Asia	>50	10–25
	Reinforced masonry		
11	Low rise	>10	2–8
12	Mid-rise	>20	15–40
13	**Confined masonry**	<10	2–6
14	Catastrophic collapse (pancake)	>60	10–40
15	Mixed	>30	15–20

Source: Modified from So 2016.

through an analytical approach, using analytical fragility models that allow different collapse mechanisms and associated collapsed debris to be estimated. By estimating the fatality risk for different types of buildings in this way, additional knowledge on the structural defects of the buildings was also obtained, which was then used to guide the strengthening recommendations to the buildings in the region.

Crowley et al.'s (2017) study takes the analytical modelling of unreinforced masonry structural collapse by other researchers, e.g. Furukawa and Ohta (2009) a step further by associating collapse with casualties. This has the potential of generating a suite of casualty fragility functions for each global collapse state under different levels of ground shaking. As discussed in the previous work by researchers (Okada and Takai 2000; So and Pomonis, 2012; So et al. 2018), this would be a significant step forward for earthquake casualty estimation.

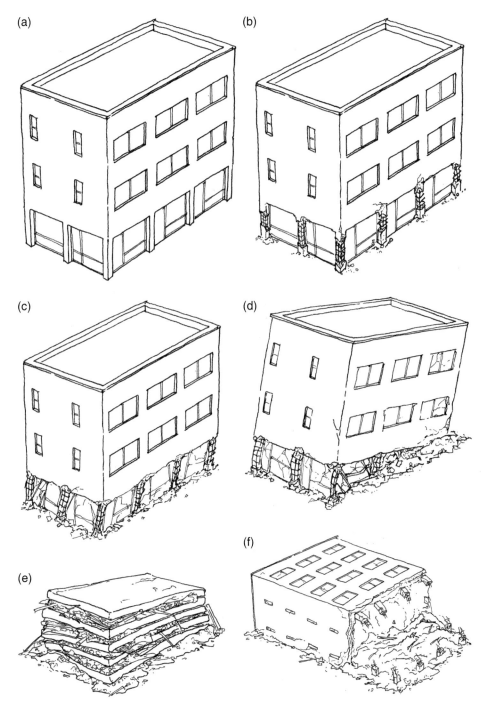

Figure 6.18 Sketches of different collapse mechanisms of a midrise reinforced concrete building, demonstrating the very different resulting IVR. (a) undamaged (b) column failure (c) soft storey deformation (d) soft storey failure (e) pancake collapse (f) overturn

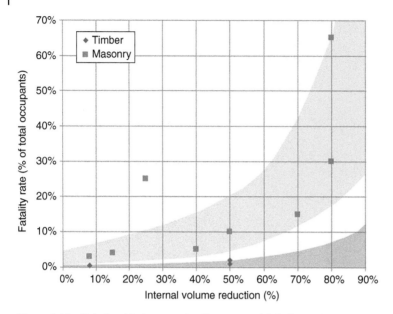

Figure 6.19 Relationship between fatality rates and IVR. *Source:* Modified from So (2016).

Volume loss characteristics:

- Mw7.6 Kocaeli earthquake in 1999, Turkey
- Collapse mechanism: pancake (all floors)
- Damage level: D5
- Estimated volume loss: 99%

- Mw7. 1 Van earthquake in 2011 , Turkey
- Collapse mechanism: soft storey ground floor
- Damage level: D5
- Estimated volume loss: 20%

Figure 6.20 Examples of photographs used for IVR analysis. *Source:* Photograph sources – left: CAR; right: EERI.

6.6.2 Estimating Fatalities Due to Falling Debris

Although acknowledged as a contributing factor to overall casualties from earthquakes (see Figure 6.2), there has been limited research on quantifying the risk of injuries and deaths to pedestrians, people trying to escape buildings and people in vehicles due to falling debris during earthquakes.

In Mw6.1 Christchurch earthquake in 2011 falling debris was responsible for the death of 35 out of 181 fatalities (Abeling et al. 2018), while in the Mw6.9 Loma Prieta earthquake in 1989, 8 of the 57 direct deaths were also due to falling debris in Santa Cruz and San Francisco.

The distance over which debris spreads is an important indicator of potential casualties to those outside the building. Abeling et al. (2018) call this the Pedestrian Hazard Area (PHA). In their paper, an equation was generated to estimate the number of casualties in the PHA, based on experiences from the Mw7.0 Darfield earthquake in 2010 and the Mw6.1 Christchurch earthquake in 2011. Parameters in the equation include the likelihood of a building to generate death or serious injury due to falling debris, the ratio of the hazard area covered in debris, the probability of a person escaping the hazard area unharmed, and probability of occurrence for the casualty state conditioned on a pedestrian being struck by falling debris. However, large uncertainties in the input parameters of the proposed model have resulted in uncertainty bounds greater than two orders of magnitude in some cases.

Recognising a need to examine the lethality potential of falling building debris and collapsing buildings onto adjacent pavements and streets, in addition to assessing building damages from photographic archives, So et al. (2018) also used the dataset discussed in the previous section to classify the amount of debris generated around the envelope of the building, and thus potentially pose a danger to those in adjacent streets or other public spaces. Using rapid visual analysis, an observation spreadsheet was created containing the following fields: maximum distance of debris from building (m); area of debris (m²); maximum depth of debris (m); main source of debris; percentage of element failed and the nature of most debris. Examples of photographs analysed for falling debris are shown in Figure 6.21.

Apart from helping to quantify the coverage of debris associated with different building types and damage grades, this study also showed the propensity of failures at lower damage grades of D3 and D4s to cause injuries and deaths, which warrants further investigations. Such studies could be used to fast track strengthening programmes of vulnerable structural and non-structural elements in masonry buildings.

Debris characteristics:

- Mw6.9 Kobe earthquake in 1995, Japan
- Max distance of Debris: <1m
- Area of debris: 10–25 m²
- Max depth of debris: <1m
- Main source of debris: supports
- Percentage of element failed: <10%,
- Nature of most debris: concrete

- Mw6.7 Erzincan earthquake in 1992, Turkey
- Max distance of Debris: 1–3m
- Area of debris: 25–50 m²
- Max depth of debris: 1–2m
- Main source of debris: brick infill
- Percentage of element failed: 10–30%
- Nature of most debris: Brick

Figure 6.21 Examples of photographs analysed for the peril of falling debris. *Source:* Photograph sources: EEFIT.

6.7 Estimating Casualties from Secondary Hazards and Cascading Effects

Landslides, avalanches, tsunami and other causes account for nearly 25–30% of global earthquake deaths since 1900 (CATDAT). Attempts to estimate casualties from these causes usually employ a statistical approach. The estimates will be based on the probability and locations of the hazards and an estimate of populations exposed to the hazards. For example, for earthquake-induced landslides, Nowicki Jessee et al. (2020) compiled a global dataset of earthquake-induced landslide fatalities and proposed a global landslide exposure index based on the exposure of population to expected landslide occurrence. Work on improving such estimates are crucial for informing mitigation policies in minimising the impact of these secondary perils. The focus of mitigation actions for landslides and tsunamis would be different from the strategies for ground motions and would include relocation away from landslide and tsunami risk zones, infrastructural defences and stabilisation measures, and warning systems.

The threat of cascading events is also significant, but difficult to model. Some notable cascading events in the past century include the Mw8.1 Great Kanto earthquake in 1923, fire which was fanned by an on-going typhoon; the Mw6.6 Niigata earthquake in 2007 also occurred during Typhoon Songda. In 1987 in Napo province, the first of a series of three earthquakes of the Mw7.2 Ecuador earthquake sequence, which happened 135 minutes before the main earthquake, triggered huge mudslides that swept down the hillsides, depositing mud, rocks, and trees in the river valleys of this Amazon region. In some places, whole mountain sides came down. The huge volume of debris formed natural dams across the rivers. The second larger quake a few hours later caused the temporary dams to collapse and loosened torrents of water, mud and debris. Many of the people in the area were swept away in the deluge, along with their homes, vehicles, and livestock. Some of the affected area was covered by as much as 10 m of mud.

The Fukushima nuclear compound disaster after the Mw9.1 Great Tohoku earthquake and tsunami in 2011 also provides a stark warning for locations with nuclear power plants in seismic areas. The inclusion of cascading scenarios in loss estimation models may not be feasible due to the number of assumptions which need to be made, but the devastating impact of a series of events certainly justifies qualitative assessments and detailed planning in earthquake-prone areas.

6.8 The Way Forward

Understanding what happens to people during an earthquake and how to prevent injuries and deaths is at the heart of earthquake protection. However, as shown in this chapter, the process of identifying and therefore modelling where the affected population will be located, how they will be impacted and behave is not simple. In our review of the current status of earthquake casualty modelling and our experience from talking to survivors of recent events, there are three main areas where we can explore and make progress in casualty loss estimation.

6.8.1 Data Collection

The lack of reliable and standardised casualty data has hindered efforts in this field. We need to make efficient use of resources to capture perishable data to improve models and identify areas for prioritised earthquake preparedness. For example, there is a wealth of information from International Search and Rescue databases and rescuers' accounts that may help supplement our understanding of survival and casualties in collapsed buildings and other failures.

There is an urgent need to improve exposure models, both in terms of the built environment, especially for non-residential buildings, and mapping the population. Improvements of building exposure models would include examining built floor areas by type of structure, number of floors, period of construction, seismic vulnerability and building use classes.

Instead of a building by building approach, profiles of earthquake-prone areas could be created. These could describe affected areas in terms of degree of urbanisation, urban complexity, rural areas, the layout of the affected settlements, for example where there are many narrow streets, and existence of vulnerable buildings with high occupancies (churches, factories, malls, old or non-seismically designed apartment blocks, etc.). In rural and mountainous areas, there will be issues with access, varied quality of housing and a probability of high percentage of building stock collapsing. This would in turn hinder SAR and survival rates but unless these anomalies are identified at the outset, they cannot be planned for.

With the increased use of smart phones and other mobile devices, the uses of mobile data to provide more accurate patterns for pedestrian, vehicle and occupancy models could be explored. The number of people located within a hazard area during an earthquake is dependent upon factors such as the day of the week, if it is a working day or a holiday, if it is an area where large population influx occurs due to commuting workers, or other visitors and tourists, the time of day, the busyness of urban streets and the building use.

There are also other factors that influence the population at risk and in some regions, the variation in population can be significant. These include the season of occurrence, especially in areas where tourism can multiply the exposed population. The season can also affect the chances of survival of people trapped under rubble. Further, cultural aspects (e.g. women at work or at home, men in construction sites or in the crop fields), and the age pyramid of the population in the affected area can all affect survival rates. More data is needed on all these aspects.

6.8.2 Rethinking Loss Models

In order to account for the variability in collapse mechanisms and its impact on casualties, future research and analytical loss estimation models should be directed to expanding the probability of collapses for different structural types. Analytical models could be calibrated with observed damage and give probabilities of collapses based on overall failure mechanisms, IVR and building use types. The influence of roof types in structural collapse and fatality rates should also be considered.

It may also be time to take this field beyond single hazard loss estimation and towards mitigation based on societal fatality risk studies. Tsang et al. (2020) argues that by comparing earthquake risk against other societal risks, one may be able to convince policy-makers that the resulting earthquake fatality risk for a society is unacceptable, based on a regulatory requirement that aims to limit the mortality rate to 'as low as reasonably practicable (ALARP)'. The concept of ALARP is central to the United Kingdom's health and safety system and is in use in Australia and New Zealand. In the United Kingdom, it is a key part of the general duties of the Health and Safety at Work, etc., Act 1974 and many sets of health and safety regulations that the Health and Safety Executive and the UK's Local Authorities enforce (Hurst et al. 2019). ALARP is also used in risk analyses globally where the regulatory framework is 'goal' setting rather than 'prescriptive' (Taylor and Israni 2014).

6.8.3 Life-Safety Designs

The primary goal of casualty estimation is to quantify human losses due to earthquakes and prompt mitigation actions. Reviewing causes of casualties from past earthquakes can help focus these efforts and help engineers 'design out' these main culprits. In developing countries, the focus may need to be shifted from an earthquake-resistant dwelling to one that provides life safety. It could be as simple as changing to a lighter weight roof.

Critical buildings such as schools, hospitals and churches have long been considered as places of mass gathering and have garnered special attention. However, as globalisation and urbanisation gather pace, working and congregation patterns have changed. Home working may become more prevalent in some countries whilst in others, overcrowded factories as witnessed in the 2013 Dhaka garment factory collapse, have become the norm. To prevent future tragic losses of life, a critical review of where and how one can provide basic life safety is key.

References

Abeling, S.R., Cutfiled, M., and Dizhur, D.D. et al. (2018). Casualty estimation from falling masonry debris based on 2010/2011 Canterbury earthquake damage. *10th International Masonry Conference*, Milan, Italy (9–11 July 2018).

Aghamohammadi, H., Mesgari, M.S., Mansourian, A., and Molaei, D. (2013). Seismic human loss estimation for an earthquake disaster using neural network. *International Journal of Environmental Science and Technology* 10 (5): 931–939. https://doi.org/10.1007/s13762-013-0281-5.

Allen, T.I., Wald, D.J., Earle, P.S. et al. (2009). *EXPO-CAT: Population Exposure to Intensity for Each Atlas ShakeMap*. U.S. Geological Survey. https://doi.org/10.5066/P9X3AXVJ.

Ceferino, L., Kiremidjian, A., and Deierlein, G. (2018). Regional multiseverity casualty estimation due to building damage following a Mw 8.8 earthquake scenario in Lima, Peru. *Earthquake Spectra* 34 (4): 1739–1761.

Coburn, A.W. and Spence, R.J.S. (2002). *Earthquake Protection*, 2e. Chichester: Wiley.

Coburn A.W., Spence R.J.S., and Pomonis, A. (1992). Factors determining casualty levels in earthquakes: mortality prediction in building collapse. *Proc of the 10th WCEE* (19–24 July 1992), Madrid, Spain.

Crowley, H., Polidoro, B., Pinho, R., and van Elk, V. (2017). Framework for developing fragility and consequence models for local personal risk. *Earthquake Spectra* 33 (4): 1325–1345.

Daniell, J.E. and Wenzel, F. (2014). The production and implementation of socioeconomic fragility functions for use in rapid worldwide earthquake loss estimation. *Proc of the 2nd ECEES*, Istanbul, Turkey (25–29 August 2014).

Daniell, J.E, Vervaeck, A., Khazai, B., and Wenzel, F. (2012). Worldwide CATDAT damaging earthquakes database in conjunction with Earthquake-report.com – presenting past and present socio-economic earthquake data. *Proceedings of the 15th World Conference of Earthquake Engineering* (24–28 September 2012), Lisbon, Portugal.

FEMA (2015). Hazus–MH 2.1: Technical Manual (accessed 22 March 2020). http://www.fema. gov/media-library-data/20130726-1820-25045-6286/hzmh2_1_eq_tm.pdf (accessed 2 October 2020).

Feng, T., Hong, Z., Fu, Q. et al. (2014). Application and prospect of a high-resolution remote sensing and geo-information system in estimating earthquake casualties. *Natural Hazards and Earth System Sciences* 14: 2165–2178. https://doi.org/10.5194/nhess-14-2165-2014.

Furukawa, A. and Ohta, Y. (2009). Failure process of masonry buildings during earthquake and associated casualty risk evaluation. *Natural Hazards* 49: 25–51.

Geohazards International (2018). Developing messages for protective actions to take during earthquake shaking. WWW Document. https://4649393f-bdef-4011-b1b6-9925d550a425. filesusr.com/ugd/08dab1_503d1009b7d9438db1a462f91a430522.pdf (accessed 3 June 2020).

de Groeve, T., Vernaccini, L., and Annunziato, A. (2006). Modelling disaster impact for the global disaster alert and coordination system. *Proceedings of the 3rd International ISCRAM Conference*. B. Van de Walle and M. Turoff, eds.) (May 2006). Newark, NJ.

Gueri, M. (1983). *Colombia: The Earthquake in Popayán. Disasters Preparedness and Mitigation*, Issue No. 17. PAHO.

Hurst, J., McIntyre, J., Tamauchi, Y. et al. (2019). A summary of the 'ALARP' principle and associated thinking. *Journal of Nuclear Science and Technology* 56 (2): 241–253. https://doi. org/10.1080/00223131.2018.1551814.

Jaiswal, K. and Wald, D.J. (2010a). An empirical model for global earthquake fatality estimation. *Earthquake Spectra* 29 (1): 1–16.

Jaiswal, K.S. and Wald, D.J. (2010b). Development of a semi-empirical loss model within the USGS Prompt Assessment of Global Earthquakes for Response (PAGER) System. *Proc. of the 9th US and 10th Canadian Conference on Earthquake Engineering: Reaching Beyond Borders* (25–29 July 2010), Toronto, Canada.

Jaiswal, K. S., Wald, D. J., and D'Ayala, D. (2011). Developing Empirical Collapse Fragility Functions for Global Building Types. Earthquake Spectra, 27, No. 3, 775–795

Jones, N., Noji, E., Smith, G., and Krimgold, F. (eds.) (1989). *Proceedings of the International Workshop on Earthquake Injury Epidemiology for Mitigation and Response*, 1989. Baltimore, MD: Johns Hopkins University.

Keefer, D. (1984). Landslides caused by earthquakes. *Geological Society of America Bulletin* 95: 406–421.

Keefer, D.K. (2002). Investigating landslides caused by earthquakes – a historical review. *Surveys in Geophysics* 23 (6): 473–510.

Kuwata, Y., Takada, S., and Bastami, M. (2005). Building damage and human casualties during the Bam-Iran earthquake. *Asian Journal of Civil Engineering (Building and Housing)* 6: 1–19.

Lallemant, D., Burton, H., Ceferino, L. et al. (2017). A framework and case study for earthquake vulnerability assessment of incrementally expanding buildings. *Earthquake Spectra* 33: 1369–1384.

Lambie, E., Wilson, T.M., Johnston, D.M. et al. (2016). Human behaviour during and immediately following earthquake shaking: developing a methodological approach for analysing video footage. *Natural Hazards* 80 (1): 249–283. https://doi.org/10.1007/s11069-015-1967-4.

Macabuag, J., Raby, A., Pomonis, A. et al. (2018). Tsunami design procedures for engineered buildings: a critical review. *Proceedings of the Institution of Civil Engineers – Civil Engineering* 171: 166–178. https://doi.org/10.1680/jcien.17.00043.

Marano, K.D., Wald, D.J., and Allen, T.I. (2010). Global earthquake casualties due to secondary effects: a quantitative analysis for improving rapid loss analyses. *Natural Hazards* 52: 319–328. https://doi.org/10.1007/s11069-009-9372-5.

Murakami, H. (1996). Human casualty due to the 1995 Great Hanshin-Awaji earthquake disaster in Japan. *11th World Conference on Earthquake Engineering,* CD-ROM Paper No. 852 (23–28 June 1996), Acapulco, Mexico, 8.

Nichols, J.M. and Beavers, J.E. (2003). Development and calibration of an earthquake fatality function. *Earthquake Spectra* 19 (3): 605–633.

Nowicki Jessee, M.A., Hamburger, M.W., Ferrara, M.R. et al. (2020). A global dataset and model of earthquake-induced landslide fatalities. *Landslides* 17: 1363–1376. https://doi.org/10.1007/s10346-020-01356-z.

OCHA (2006). Indonesia earthquake response plan 2006. http://www.unocha.org/sites/dms/CAP/Revision_2006_Indonesia_Earthquake_RP.doc.

Ohta, Y. and Ohashi, H. (1985). Field survey on occupant behavior in an earthquake. *International Journal of Mass Emergencies and Disasters* 3: 147–160.

Okada, S. (1996a). Study on the evaluation of seismic casualty risk potential in dwellings: part 2, human behavior to evacuate during the 1993 Kushiro-oki earthquake. *Journal of Structural and Construction Engineering* 481: 27–36. (in Japanese).

Okada S. (1996b). Description of indoor space damage degree of building in earthquake. *11th World Conference on Earthquake Engineering,* 3/4 (CD-ROM) Paper No. 1760 (23–28 June 1996), Acapulco, Mexico.

Okada, S. and Takai, N. (1999). Classifications of structural types and damage patterns buildings for earthquake field investigation. *Journal of Structural and Construction Engineering, Transactions of the Architectural Institute of Japan* (524): 65–75. (in Japanese).

Okada, S. and Takai, N. (2000). Classifications of structural types and damage patterns of buildings for earthquake field investigation. *Proceedings of the 12th World Conference on Earthquake Engineering,* Auckland, New Zealand. Paper no.705.

Pan, S.T., Cheng, Y.Y., Wu, C.L. et al. (2019). Association of injury pattern and entrapment location inside damaged buildings in the 2016 Taiwan earthquake. *Journal of the Formosan Medical Association* 118 (1 Pt 2): 311–323. https://doi.org/10.1016/j.jfma.2018.05.012. Epub 2018 May 30. PMID: 29857951.

Petal, M.A. (2004). *Urban Disaster Mitigation and Preparedness: The 1999 Kocaeli Earthquake.* Los Angeles: *Diss. Univ. of California.*

Pomonis, A. (2005). *Estimating Casualty Rates due to Earthquake Ground Shaking in Japan.* RMS Inc.

Pomonis, A., So, E., and Cousins, J., (2011). Assessment of fatalities from the Christchurch New Zealand earthquake of February 22nd 2011. Seismological Society of America 2011 Annual Meeting, Memphis, TN (13–15 April).

Reinoso, E., Jaimes, M.A., and Esteva, L. (2017). Estimation of life vulnerability inside buildings during earthquakes. *Structure and Infrastructure Engineering* 14 (8): 1140–1152. https://doi.org/10.1080/15732479.2017.1401097.

Rossetto T, Raby A, Brennan A et al. (2019). The Central Sulawesi, Indonesia Earthquake and Tsunami of 28th September 2018 – A Field Report by EEFIT-TDMRC.

Samardjieva, E. and Badal, J. (2002). Estimation of the expected number of casualties caused by strong earthquakes. *Bulletin of the Seismological Society of America* 92: 2310–2322.

Seligson H.A., Shoaf K.I., and Kano, M. (2006). Development of casualty models for non-ductile concrete frame structures for use in PEER's performance-based earthquake engineering framework. *Proceedings of 100th Anniversary Earthquake Conference Commemorating the 1906 San Francisco Earthquake/8th U.S. National Conference on Earthquake Engineering*, San Francisco, CA (18–22 April 2006).

Shiono, K. (1995). Interpretation and published data of the 1976 Tangshan, China earthquake for the determination of a fatality rate function. *Earthquake Engineering* 11 (4): 155–163.

Shoaf, K.I., Sareen, H.R., Nguyen, L.H. et al. (1998). Injuries as a result of California earthquakes in the past decade. *Disasters* 22 (3): 218–235.

So, E.K.M. (2009). The assessment of casualties for earthquake loss estimation. PhD Dissertation. University of Cambridge, UK.

So, E. (2016). Estimating fatality rates for earthquake loss models. In: *Springer Briefs in Earth Sciences*. Springer. ISBN: 978-3-319-26837-8, 62 p.

So E.K.M. and Pomonis A. (2012). Derivation of globally applicable casualty rates for use in earthquake loss estimation models. *Proceedings of 15th World Conference on Earthquake Engineering*, Paper 1164 (24–28 September 2012), Lisbon, Portugal.

So, E., Baker, H., and Spence R. (2018). Estimation of casualties caused by building collapse in earthquakes through assessment of volume loss and external debris spread. *Proceedings of 16th European Conference on Earthquake Engineering-Thessaloniki 2018* (18–21 June 2018), Thessaloniki, Greece.

Spence, R. and So, E. (2013). Estimating shaking-induced casualties and building damage for global earthquake events: a proposed modelling approach. *Bulletin of Earthquake Engineering* 11: 347–363.

Taylor, M. and Israni, C. (2014). *Understanding the ALARP Concept: Its Origin and Application*. Society of Petroleum Engineers. https://doi.org/10.2118/168486-MS.

Tsang, H.H., Daniell, J.E., Wenzel, F. et al. (2020). A universal approach for evaluating earthquake safety level based on societal fatality risk. *Bulletin of Earthquake Engineering* 18: 273–296. https://doi.org/10.1007/s10518-019-00727-9.

Urrutia, J.D., Bautista, L.A., and Baccary, E.B. (2014). Mathematical models for estimating earthquake casualties and damage cost through regression analysis using matrices. *Journal of Physics: Conference Series* 495: 1–15.

Wyss, M. (2004). Real-time prediction of earthquake casualties. In: *Disasters and Society – From Hazard Assessment to Risk Reduction* (eds. D. Malzahn and T. Plapp), 165–173. Karlsruhe: Logos.

Yin, Y., Wang, F., and Sun, P. (2009). Landslide hazards triggered by the 2008 Wenchuan earthquake, Sichuan, China. *Landslides* 6: 139–152.

7

How Can Buildings Be Improved?

7.1 Introduction

This chapter considers how the earthquake-resistance of buildings worldwide can be improved in order to reduce and eventually eliminate the damage and casualties caused by earthquakes.

In those countries which have been successful in reducing earthquake impacts over the last century, it has been found that where buildings are designed by experienced engineers and making use of standards or codes of practice developed from the experience of past events, damage can be contained. In California, Japan, New Zealand and Chile damage caused by ground shaking has been relatively modest in recent events (see Chapter 8). All of these countries not only have codes of practice which are more or less universally implemented; they also have an engineering profession well-educated in the interpretation of the codes. Those buildings which did collapse have in most cases been found to have been built before the current generation of earthquake codes was adopted (usually in the 1970s). This chapter first (Section 7.2) considers the design of engineered buildings and the codes of practice they are based on. It reviews the development of the codes and what current codes contain. It discusses the training of engineers, and the framework of legislation and building control needed for codes to be implemented. And it identifies some limitations of existing codes.

However, the implementation of codes of practice in all new buildings is not enough to eliminate or reduce earthquake damage, when much of the existing building stock in many earthquake-prone countries was built before the current codes were in force. This is particularly a problem in a number of the high-income countries, especially in Europe, where the rate of replacement of old building stock is slow. It is also a problem in countries where codes exist but have been poorly implemented in the past, such as Turkey and China. Not only homes and apartment buildings, but massive numbers of public buildings too, are built in ways which are now known to be unsafe in the event of expected levels of ground shaking. School buildings in particular have a very poor record of performance in earthquakes. Section 7.3 therefore considers how existing unsafe buildings can be improved. It discusses the economics of strengthening as opposed to replacement. It considers the problems associated with implementing strengthening programmes. And it gives some examples of programmes of assessment and strengthening of particular classes of buildings carried out in different countries. Details of strengthening techniques are given in Chapter 5.

Why Do Buildings Collapse in Earthquakes?: Building for Safety in Seismic Areas,
First Edition. Robin Spence and Emily So.
© 2021 John Wiley & Sons Ltd. Published 2021 by John Wiley & Sons Ltd.

Implementation of codes has been a successful approach to risk reduction largely in high-income countries. For most of the population of the world's poorest earthquake-prone countries, neither codes of practice nor official building strengthening programmes will be fully effective in reducing future death tolls or earthquake damage. As seen in Chapter 3, most of the inhabitants of the poor countries, whether in urban or rural areas, live in non-engineered buildings which are built with little consideration for earthquakes, and frequently using poor-quality materials. The builders of these buildings are often the owners themselves, generally assisted by local artisan builders. They are not reached by codes of practice or by any building control process. This is the most intractable aspect of the world's earthquake problem. Efforts to reach these populations need to be through programmes targeted at the house-owners and artisan builders, and through builder-training programmes supported by simplified guidance documents which explain the techniques to build more safely. Section 7.4 describes some of the programmes which have been undertaken in countries such as Indonesia, India and Nepal for improving non-engineered buildings and argues the need for a massive expansion of such programmes.

An essential basis for any successful programme of improving earthquake protection is a society which is fully aware of the earthquake risk, and individuals who know about the nature of earthquakes, how to behave when an earthquake occurs, and what action they can take to make their homes, workplaces and communities safer. Public awareness programmes take different forms in different countries. Section 7.5 describes recent public awareness programmes in several countries and considers how effective they have been.

7.2 Design of Engineered Buildings

7.2.1 The Development of Earthquake Codes

Early codes were developed following the experience of particular earthquakes and were based on the banning of types of construction which performed particularly badly. The concept of design for a lateral force was first used in Italy after the M7 1908 Messina earthquake, when a government commission recommended design for a lateral load of 1/12th (later 1/8th) of the weight supported. Similar rules were adopted in Japan after the Mw8.1 Great Kanto earthquake in 1923, and in California after the Mw6.4 Long Beach earthquake in 1933. Since that time earthquake, codes have been in a constant state of development. Codes have been able to build on a growing understanding of the location of active earthquake faults and the likely magnitudes and frequencies of earthquakes in different regions. They have also made use of developing research on the behaviour of structures under earthquake loading, as well as knowledge acquired from the observed performance of buildings in successive earthquakes.

An important aspect of earthquake codes is that they do not intend that structures should be capable to resisting their design loads within the elastic range of response of the structure. Some inelastic behaviour is permitted, leading to some localised damage, so long as the building can resist the maximum expected earthquake load without life-threatening damage or structural collapse. Code drafting committees have taken the view that, given the rarity of large earthquake shaking in any location, designing buildings to resist without any damage would be economically impractical (Hamburger 2003). The principal aim has been life safety.

Codes now in use derive, essentially, from three sources (Hamburger 2003). First, they derive from the experience of the observation of real structures in earthquakes, each major event giving rise to additional elements of the codes. Second, they derive from a growing theoretical understanding of earthquake loading on structures and its consequences, developed in academic institutions and backed by increasingly sophisticated testing programmes, both in the laboratory and in actual buildings. But third, they are also based on engineering judgement. Structural engineers in each affected region have worked collectively to develop codes which are not only safe for individual buildings and their occupants but also applicable by the profession and the building industry, and politically acceptable in terms of their performance objectives.

Over time, separate sets of codes have been developed in the USA (BSSC 1997), in Japan (Building Standards Law of Japan), in New Zealand (New Zealand Building Standards Law) and in Europe (CEN 2004). Although there are many detailed differences in terminology and in the detail of procedures, (a recent review (Booth 2018) identified 36 national codes), each is based on one of these leading codes.

Codes are of course not static; they have developed over time and are still developing. In New Zealand, following the experience of the 2011 Christchurch earthquake in which many buildings with repairable damage were nevertheless demolished for economic reasons, the concept of 'damage limitation', has been developed and will in the future be a basis for design. Likewise, in California the concept of resilient design has been adopted by the leading cities of Los Angeles and San Francisco (see Chapter 8). Peter Fajfar, writing about the future of codes in Europe, has proposed that an explicitly probabilistic risk-based approach to the design of structures should in the future be allowed for in the Eurocode (Fajfar 2018). And Edmund Booth (2018) has suggested that the process of development of codes needs in future to become more inclusive of the wider community of users of standards, and those affected by them, rather than being 'colonised' by the technical community as at present.

The cost of making new buildings more seismically resilient is often seen as a limitation to the development of codes aiming to reduce damage in future earthquakes. This is not as severe a problem as might be expected. Studies have shown that, in the USA, the additional cost of construction to make a building which will resist a 50% increase in earthquake loading, and be stiffer, can be expected to add only 1% to construction costs and perhaps only 0.5% to the purchase price of a building (Porter 2019). And studies have repeatedly shown that the saved cost of loss and repair following a damaging event can be many times greater than the additional cost of construction to a higher standard of resilience would have been (Moullier and Krimgold 2015).

7.2.2 Performance Objectives

The concept of life safety is straightforward: in the event of the largest considered earthquake, a building may be damaged, but will not collapse. But other performance objectives may be important, particularly for buildings which need to maintain their function following an earthquake. The current revision of Eurocode 8 (still in draft) (CEN 2020) defines four separate performance objectives as follows:

- Fully operational (OP). Slight damage, but allowing continuous operation of the building's functions.
- Damage limitation (DL). Slight structural damage; the cost of non-structural damage is small compared with the value of the structure and structural resistance is unimpaired.

Table 7.1 Performance objectives set out in the new revision of Eurocode 8.

Performance objective	Return period for ordinary buildings (Class CC2) (years)	Return period for buildings of community importance (Class CC3-a) (years)	Return period for buildings essential for civil protection (Class CC3-b) (years)
Damage limitation (DL)	60	60	100
Significant damage (SD)	475	800	1600
Near collapse (NC)	1600	2500	5000

Source: Modified from CEN (2020).

- Significant damage (SD). Structure is significantly damaged with moderate permanent storey deformations, but retains its vertical load-bearing capacity and a residual lateral strength and stiffness. Non-structural components are damaged, but partitions and infills have not failed out-of-plane.
- Near collapse (NC). Structure sustains heavy damage with large permanent storey deformations, but building retains its vertical load-bearing capacity.

The new revision of Eurocode 8 also associates to the last three performance objective the return period of the earthquake for which the structure should be designed, distinguishing between the objectives required for ordinary buildings (consequence class CC2), those required for buildings of community importance (e.g. schools, assembly halls, consequence class CC3-a) and buildings essential for civil protection such as hospitals and fire stations (consequence class CC3-b) as shown in Table 7.1 (CEN 2020).

In practice, design of buildings is largely based on the significant damage performance objective, with other objectives checked as required in national regulations.

7.2.3 Typical Code Requirements

Most codes include, as the basic methodology, an equivalent static design procedure whereby a minimum required lateral strength is calculated. The corresponding shear force at the base of the building is then applied to the structure as a set of lateral forces applied at each storey level up the height of the building. This procedure is permitted for low-rise or medium-rise buildings which do not have significant structural irregularities, but more complex procedures are needed in other cases.

The required base shear force to design for is derived from the design peak ground acceleration (PGA) or one second spectral acceleration (SA) for the reference return period (usually 475 years, see Table 7.1). In some codes, this value can be taken from maps; or the country may be divided into a set of seismic zones for each of which a seismic coefficient is defined. In some areas, different design earthquakes are required depending on whether the earthquake is a distant, large magnitude event or a smaller more local event. A further modification of the base shear force may need to be applied according to the soil type, since soft soils tend to amplify the ground motions which occur in the underlying rocks, as discussed in Chapter 4. The soil type may be determined by published maps, or is preferably determined by a survey of the site, and an amplification coefficient is applied accordingly. To allow for a better level of protection for buildings of particular importance, such as hospitals and buildings needed in emergency operations, a further importance factor may be used to enhance the lateral force designed for.

Two further factors are generally included in the calculation. The first is a 'structural factor', which allows for the ductility of the structure, and acknowledges that during the transient force of an earthquake the structure may safely exceed the loading which would be considered safe in steady loading conditions. The second factor is the building's flexibility, measured by its 'fundamental period', the time required for a single oscillation if the building is given a lateral jolt. The more flexible the building, the lower the base shear force it will generally feel in an earthquake. The extent of this effect depends on the frequency distribution of the shaking assumed in the design earthquake.

The base shear force derived from this calculation is then applied to the building, with different loads applied at different heights, taking account of the way in which the building sways, and distributed to each structural member, columns, walls, beams and floor slabs.

The code (or associated documents) then also specify the strength and deformation properties of each structural member, according to its size and the material used, to enable the calculated load to be compared with the available strength and ductility requirements. And lateral deformations of each storey must be calculated and compared with specified maximum deflections. However, in capacity design, the strength of some elements (beams) may be governed by the requirement that they yield before the columns reach their capacity, to avoid a catastrophic collapse mechanism (Booth 2014).

Finally, the detailing of reinforcement and connections needs to be specified according the requirements of the particular code, to ensure that the structure will have adequate ductility.

Codes also typically specify design procedures for foundations, for fixing non-structural elements (cladding, services, etc.), and for securing building contents.

The design procedure outlined here is the simplest available within the codes. For more complex or multi-storey structures, more detailed analytical procedures are available, taking account of the dynamic, rather than static, nature of the actual ground motion. For certain classes of buildings (e.g. tall buildings) such procedures may be required (Booth 2014).

7.2.4 Simplified Design Rules

Separately from the formal codes, for some common structural types in certain countries, a set of deemed-to-satisfy requirements may be allowed, to enable simpler design procedures without the need for experienced professionals. For timber-frame single-family houses both in Japan and in California, such rules are available. In California, for example, where most housing is designed by developers, timber structures that are one or two stories in height can use prescriptive code requirements that define the size and quality of all materials and wall-sized framing members relative to spans and loads, and the connections needed, without the need for structural analysis (Arnold 2012).

In Nepal, recognising that it is not practical to insist that all small buildings are designed by an engineer, the 1994 code allows the use of mandatory 'rules of thumb' for certain classes of building not exceeding height and size criteria. Sets of rules are available for both load-bearing masonry construction, and for reinforced concrete buildings, the most common construction types in the country, and detailed guidance documents have been published (Government of Nepal 1994a).

7.2.5 Limitations of Codes

It is clear from the very brief overview of the code provisions given above that the use of an earthquake resistant design code for the design of structures in seismic zones is not a straightforward matter. Even the simplest procedure requires a sophisticated understanding of building structural performance in earthquakes, demanding a high level of education in engineering; and engineering judgement is needed for the selection of the correct procedure, its application to the building being designed, and the proper selection of the values for the various elements of the calculation. Beyond this, as Edmund Booth (2014) has put it:

> the use of codes should not be a substitute for sound engineering judgement. Codes describe minimum rules for standard conditions, and cannot cover every eventuality. Buildings respond to ground shaking in strict accordance with the laws of physics, not in accordance with rules laid down by a (sometimes fallible) code drafting committee.

Engineers designing buildings in earthquake areas have a very serious responsibility. As well as an understanding of the codes, they need a good education followed by experience, leading to sound judgement. And at times, they need the authority to dissuade building owners or architects from the use of complex shapes which will make the earthquake forces more difficult to deal with by the structure.

It is also important to note that the safety of the resulting building does not just depend on the codes being applied in the design office, but the resulting designs must be successfully communicated to the builders on site. The engineer has a responsibility to ensure that the design is understandable and buildable. Also, it is the responsibility of the engineer to require that adequate supervision and inspection regimes are in place to ensure that the building is built according to the design.

Thus, though this is an approach which has been successful in the right context, the application of engineering codes of practice depends on a well-developed regulatory system under which codes are seen as compulsory and are implemented, a well-educated and professionally organised structural engineering profession, and an effective system of onsite supervision and inspection. As shown by our country-by-country success and failures survey reported in Chapter 8, these conditions are largely met in the high-income earthquake-risk countries, the USA (especially California), Japan, Chile and New Zealand, and also in many of the European countries. However, in the low and middle-income countries where the world's greatest earthquake risk is concentrated (as indicated in Chapters 1 and 2), such as China, India, Indonesia and Nepal, one or more of these requirements is missing. We therefore need to consider how to improve the regulatory framework for these countries.

7.2.6 Failure of Codes in Developing Countries

In an important 2015 study, Thomas Moullier and Fred Krimgold at the World Bank's Global Facility for Disaster Risk Reduction (GFDRR) have identified a series of significant failures of codes and regulations in mitigating the effects of natural disasters, including earthquakes, in the growing cities of the developing world, and they have formulated some recommendations for improvements (Moullier and Krimgold 2015). Their report first details the reasons why codes have not been effective, based on detailed studies carried out in many countries. Principal among these reasons is the effect of uncontrolled urban expansion, which puts most of the new urban migrants beyond the reach of any building control. But the formulation of such building codes as do exist is another weakness. The report finds that codes are in most cases not adequately supported by a framework of legislation at a national level to ensure their implementation. The codes themselves are inadequate in many respects: they set unrealistic standards for construction by poor families; they fail to take account of the vernacular building techniques and local materials used by these families, and they fail to provide guidance for improved resilience of traditional forms of construction. They also fail to address the prevalent pattern of incremental construction whereby people build houses section by section over time as a means of saving. And the documents themselves are written in a language which is difficult to understand, without simplified guidance or explanatory material.

Beyond the codes, there is also a widespread failure of administration and implementation. This is partly due to an inability to recruit qualified staff for inspections. But a major factor is corruption, and a widespread practice of paying officials to acquire construction permits. The report gives the example of Turkey, a country in which good earthquake design codes exist, and yet in the Mw7.6 Kocaeli earthquake in 1999 huge numbers of recently built reinforced concrete buildings collapsed, killing 17 000 people, largely through failure of code implementation. Many other such examples are given.

The GFDRR report concludes with a set of recommendations, the most important of which are to

- Establish a sound legislative and administrative structure at the national level.
- Develop a building code suitable to local social and economic conditions that facilitates safe use of local building materials and practices.
- Strengthen implementation of the building code through plan review, site inspection and permitting at the local level.
- Provide advisory services in addition to inspection and enforcement to support code compliance.

Thus, a mixture of both top-down and bottom-up approaches to improving code effectiveness is suggested.

Finally, the report proposes setting up a programme of support for this agenda through the World Bank, in support of the 2015 Sendai Framework for Action Agenda. The programme has been in operation since 2018, and initial national action plans have been established in several countries (e.g. Jamaica, Sri Lanka). Work along similar lines is in progress in several other countries, some of which is described in later sections of this chapter.

7.3 Strengthening Existing Buildings

In parts of the world where current earthquake codes and standards are generally well-applied, the safety of the older less earthquake-resistant construction becomes a matter of increasing concern. Damage and fatalities in earthquakes is often concentrated in the older construction. It has been suggested that this problem will diminish with time as the older building stock is replaced, and this may be true of rapidly developing cities such as Istanbul with a high rate of replacement of the building stock. But in many affluent countries, the rate of replacement of the building stock is very slow. A recent report from the OECD (2019) indicates that most of the OECD countries had an annual increment to the housing stock less than 1.5% of the existing housing stock, and this includes many earthquake risk countries. Among the earthquake-prone countries providing data, in Portugal and Spain the annual increment was less than 0.5%, while in the United States it was just under 1%. In Japan, Chile and New Zealand the increment was around 1.5%. At these rates of new construction, it will clearly take a long time for the earthquake risk to reduce through the application of codes only to new construction. Indeed, rather than being taken out of use, the older building stock is often modified in ways which increase its earthquake vulnerability. And lack of maintenance and general decay as well as minor damage from previous earthquakes will also reduce the resistance of the older building stock to future earthquakes.

For all these reasons, strengthening of existing buildings (retrofitting) is assuming increasing importance in earthquake regions. Much research has been devoted to developing retrofitting techniques suitable for the different types of construction found in the building stocks of the earthquake-prone countries, which have been briefly described in Chapter 5. There is significant evidence too, of the effectiveness of strengthening, based on the relative performance of strengthened and unstrengthened buildings in subsequent earthquakes (Deppe 1988; Spence et al. 2000). Where it has been used, retrofitting has been generally found effective in reducing damage, and largely avoiding collapse, though it does not altogether eliminate the possibility of damage.

As outlined in Chapter 5, the method of retrofitting to adopt will depend on the level of earthquake resistance aimed for, its effect on the appearance of the building and the length of time the building will need to be unoccupied, as well as the details of the existing structural arrangements. In New Zealand, the Wellington City Council (Charleson 2008) lists the following benefits to encourage building owners to adopt high retrofit standards:

- Improved safety for occupants and the public
- Allowance for a change in use to increase rentable value
- Leverage for improved insurance
- Reduction in the level of damage to the building and to adjacent buildings
- Lessen impacts on business

However, there are serious difficulties in creating successful large-scale programmes for retrofitting. First, it is expensive. Typically, strengthening costs range from 20 to 40% of the cost of the replacement of the building with a new one. Where cost–benefit studies have been carried out based on known levels of earthquake risk, it is difficult to justify this expenditure to reduce the expected cost of future damage (D'Ayala et al. 1997; Smyth et al. 2004). Strengthening generally involves substantial internal structural alterations, so the cost has to include the displacement of occupants for a substantial period of time. And where apartment blocks are to be strengthened, all dwelling owners have to collectively agree to the work involved.

Governments have been reluctant to enact legislation requiring homeowners to retrofit their buildings. In 2003, following the Mw5.8 Molise earthquake in 2002 which caused major damage in Italy, an effort by the European Association for Earthquake Engineering to require general strengthening was launched through a question in the European Parliament by Mihail Papagiannakis (MEP), asking

> could the Commission formulate a Directive requiring the member states to establish programmes of assessment (according to newly formulated European Standards) of all buildings and structures in earthquake-prone areas, and strengthening those found to be inadequate?.

This request was not accepted by the Commission. No reasons were given in formal communications. However, it is known that the EU, along with governments of other advanced economies, increasingly prefers non-regulatory approaches to achieving social and environmental objectives (Spence 2004), such as education, research, voluntary codes and tax incentives, and such approaches are being adopted in a number of countries, as detailed in the survey in Chapter 8.

Thus, while there has been some owner-driven retrofitting of individual buildings in recent years, such larger-scale programmes of retrofitting as have been established and successfully accomplished in recent decades have generally focussed on public sector buildings, particularly schools, as discussed below.

7.3.1 Strengthening School Buildings

There can be little disagreement that the protection of schoolchildren needs to be given priority in any government programme for earthquake protection, and this requires attention to the safety of school buildings. Not only are children the future of any society, but while at school, they are the responsibility of those, mostly government, agencies in whose care they have been put. The governments of many countries, spurred by internationally agreed development goals, have made primary education compulsory, further enhancing the responsibility of the state to provide a safe environment for their learning. And beyond the protection of the students and their teachers, schools occupy a very important place in the life of many communities, acting as a place of refuge and community organisation in the event of a disaster, and potentially, in their construction, a model for how other buildings can be safely built.

And yet, school buildings have a dreadful record of damage and collapse in past earthquakes. This was documented in 2015 by Marla Petal, Ilan Kelman and colleagues (Petal et al. 2015). Table 7.2, showing the collapses of schools which occurred causing mass casualties in the years 2001–2015, is derived from their data.

But also important to note is that in 15 further earthquakes in the same years, many other schools collapsed or were seriously damaged, but no casualties occurred because the earthquake happened when the schools were unoccupied. All of these would also have resulted in casualties had the earthquake occurred at a different time.

Table 7.2 School collapses in earthquakes yielding multiple casualties 2001–2015.

Date/local time	Location	Magnitude (Mw)	Impact on schools	Consequences for children
12 January 2010 16.53	Port-au-Prince, Haiti	7.0	Ministry estimates 4992 schools affected	Deaths and injuries unknown. Many children with disabling injuries
12 May 2008 14.28	Wenchuan, China	7.9	175 schools and 7000 classrooms destroyed	>5300 school children died in dozens of schools
8 Oct 2005 08.50	Kashmir, Pakistan	7.6	>10000 schools collapsed	>18000 school children died and >50000 seriously injured
1 May 2003 03.20	Bingöl, Turkey	6.4	4 school buildings collapsed	84 students killed in dormitory
24 February 2003 10.03	Bachu, Xinjiang China	6.3	900 classrooms collapsed	>20 students killed in one middle school collapse
31 October 2002 11.40	Molise, Italy	5.8	San Giuliano di Pugliese infant school collapsed	26 children and 3 adults killed
26 January 2001 08.16	Gujarat, India	7.7	1884 school buildings collapsed, 5950 classroom destroyed	971 children and 31 teachers killed in school activities; 1051 students and 95 teachers seriously injured

Source: Modified from Petal et al. (2015).

Based on the evidence available at that date, Wisner et al. estimated in 2004 (Wisner et al. 2004), using reasonable assumptions, that over 5000 deaths of schoolchildren could be expected in the next decade. Tragically, this estimate was exceeded by a factor of four in the deaths which did occur, mainly in the Kashmir 2005 and Wenchuan 2008 earthquakes. These two events in particular have spurred a significant increase in school safety programmes in many countries (Petal et al. 2015), supported by international agencies.

7.3.2 OECD Programme: Scope and Principles

Following the Mw5.8 2002 Molise earthquake in Italy and the Mw6.4 2003 Bingöl earthquake in Turkey, both of which caused schools to collapse causing multiple casualties, the OECD's Programme for Educational Buildings assembled an expert group for a three-day meeting to formulate new recommendations for ensuring school safety in the OECD countries, 36 of the world's higher-income countries. The OECD's recommendation (OECD 2004) states its rationale as follows:

All too frequently strong earthquakes strike OECD member countries causing the collapse of school buildings and the death of innocent children. Although earthquakes are natural and unavoidable events, school buildings need not collapse in earthquakes. The knowledge currently exists to significantly lower the seismic risk of schools and to help prevent further injury and death of school occupants during earthquakes. The expert group finds it unconscionable that schools built world-wide routinely collapse in earthquakes due to avoidable errors in design and construction, causing predictable unacceptable and tragic loss of life.

The report set out recommendations for member governments for programmes to improve the seismic safety of schools, which include risk reduction requirements for existing as well as new facilities. Additionally, the report included detailed recommendations on creating a national seismic safety policy, on accountability, on enforcing building codes and on training and community participation. This report became an official OECD Recommendation to the governments of member countries and was adopted by more than 20 out of the 36 OECD countries. Programmes for strengthening existing buildings were initiated or accelerated as a result this recommendation in a number of the OECD countries, including Italy, Portugal, Greece and Turkey, though by no means all have adopted or acted on it.

7.3.3 World Bank Programme; Scope and Principles

The World Bank's Global Programme for Safer Schools (GPSS) was launched in 2014, responding to the UN's internationally adopted 2013 Comprehensive School Safety (CSS) framework. GPSS aims to boost and facilitate informed and large-scale investments in the safety and resilience of new and existing school infrastructure at risk from all kinds of natural hazards, including earthquakes, floods and windstorms. Its focus is on public school infrastructure in developing countries (http://www.worldbank.org/en/topic/disasterriskmanagement/brief/global-program-for-safer-schools.print).

GPSS specific aims include (World Bank 2017) to

- Build evidence-based knowledge and tools for informing investments at scale and long-term engagement on safer schools;
- Play an important role in promoting and facilitating the integration of risk reduction criteria into World Bank-funded education infrastructure projects; and
- Integrate safety improvements into comprehensive solutions for quality learning environment for children in developing countries.

Since 2014, the GPSS programme has initiated activities in 12, mainly low-income, countries including Nepal, Indonesia, El Salvador and Turkey, to name those most at risk from earthquakes. Other countries have their own independent programmes.

7.3.4 Global Progress in Achieving School Safety: Some Examples

The last decades have thus seen the beginnings of a global programme to create safer schools in earthquake areas. But while a few have achieved a high degree of success, in many countries, these programmes have been progressing very slowly. The following sections give a brief account of the programmes in some specific cities or regions.

7.3.4.1 California

In California's Mw6.4 Long Beach earthquake of 1933, 70 schools were destroyed and 120 schools suffered major damage. One month later the State of California enacted the then-innovative Field Act, governing the construction of new schools and requiring high standards of earthquake protection. Since that date, new public schools have been designed to contemporary standards or better. Moreover, in the late 1960s, new regulations required pre-Field Act buildings to be retrofitted or demolished (California Seismic Safety Commission 2004). And school buildings built to pre-1978 standards have in recent years been reviewed and retrofitted where needed. Only some older private school buildings (housing about 10% of California's school enrolment) may still need assessment. But in general, the schools in California are as safe as anywhere in the world.

7.3.4.2 Japan

The Japanese national government programme for making schools earthquake-resistant was introduced in 1978, but was initially focussed in the Tokai and southern Kanto regions where the earthquake risk was considered highest. But the Mw6.9 Kobe (Great Hanshin-Awaji) earthquake in 1995 affected another region and damaged over 3800 schools, and as a result, subsidies for seismic retrofitting were extended to the entire nation. However, surveys conducted in 2002 showed that local governments were responding very slowly. By that date, only 31% of all buildings built using pre-1981 standards had been assessed, and only 44.5% were confirmed as earthquake-resistant. New guidelines for the promotion of earthquake-resistant school buildings were issued in 2003, including timetables for compliance. As a result of this programme, the rate of seismic assessment and retrofitting increased dramatically, and by 2015, 95.6% of schools were considered earthquake-resistant, with the programme expected to be completed by 2016 (World Bank 2016). Since 2003 approximately 52000 buildings have been confirmed earthquake-resistant either through seismic diagnosis or by retrofitting or reconstruction. As in California, promoting seismic retrofitting in private schools remains a challenge, and further challenges seen in Japan are making schools safer from other hazards (windstorm, volcanic eruption), making non-structural elements of school buildings hazard-resistant, and dealing with the inevitable increase in vulnerability of school buildings through ageing.

7.3.4.3 Turkey

Following the disastrous Mw7.6 Kocaeli earthquake in 1999 and spurred also by seismological studies which showed a very high risk of an earthquake affecting Istanbul within 30 years (Parsons et al. 2000), the Istanbul Provincial Government put into place a series of studies leading to the formulation in 2003 of the Earthquake Master Plan for Istanbul (Bogazici University et al. 2004). An important element of this plan is the inspection and strengthening of public buildings, including schools and dormitories, hospitals and administrative buildings. Over the last decade, this programme has been implemented through the ISMEP project, started in 2006 as a partnership between the Istanbul Provincial Government and the World Bank. Through ISMEP, between 2006 and 2013, around 2500

Figure 7.1 Retrofitted elementary school in Istanbul under ISMEP programme. *Source:* Photo: ISMEP.

buildings were identified as priorities for investigation, of which over 840 were targeted for urgent action. This included nearly 600 school buildings (ISMEP 2014) (Figure 7.1).

According to a more recent report (IPKB 2020), under ISMEP, Istanbul has 'retrofitted and/or reconstructed 1135 school buildings, ensuring an education in more secure buildings for over 1.5 million students by reinforcing approximately 4 million square metres of school buildings against earthquakes'.

7.3.4.4 Colombia

In the Mw6.1 Quindio earthquake of 1999, 74% of the schools in the Colombian city of Armenia were destroyed. Fortunately, the earthquake occurred in the school vacation, so the more than 9000 students who might have been in those schools were unharmed. But the event raised awareness of the risk to schoolchildren in all the major cities of Colombia. In 2002, the City of Bogota carried out a study of the earthquake risk in its public schools. It found that most of the city's 710 schools (comprising more than 2800 buildings) were constructed in the 1960s without considering earthquake requirements. In 2004, with encouragement from the World Bank, Bogota's administration promoted a programme to improve, rehabilitate and retrofit 201 of the highest priority buildings, through an investment of US$460mn of its own resources. Colombia has for some decades had a National System for Disaster Risk Management, which requires seismic vulnerability evaluation and rehabilitation of essential buildings, with an extended time-limit for compliance. However, beyond Bogota, seismic retrofitting of schools has been a slow process, with no action at all in some places, especially the small towns (Cardona et al. 2015). There remains a huge and growing stock of public buildings, and also many private schools, in need of strengthening, for which funds and political commitment are not yet available.

7.3.4.5 Nepal

School buildings in much of Nepal are built using traditional materials such as earth and stone and using only the skills of local craftsmen, and such buildings are highly vulnerable to earthquakes. A survey conducted by the National Society for Earthquake Technology (NSET) in 2000 of 900 public school buildings in the Kathmandu Valley, found that none

Figure 7.2 School building in Nepal, before and after strengthening, 1998. *Source:* Photo: Bothara et al. (2018). Reproduced with permission of NSET.

of them were earthquake-resistant, and more than 60% were constructed of weak masonry materials. As a result NSET initiated its School Earthquake Safety Programme (SESP) in 2000, with three components – training of masons, training of students, teachers and parents on earthquake preparedness planning, and seismic retrofit or reconstruction of schools. With international development aid support from USAID, Asian Development Bank and the World Bank, the SESP has been operational for more than 20 years. In that time (NSET 2017), it has retrofitted 300 school buildings (Figure 7.2), provided training to over 200 engineers, 2000 masons and 11 000 teachers.

Importantly, in the Mw7.8 Gorkha earthquake of 2015 which caused devastation over much of the country with the loss of more than 22 000 schools, none of the schools retrofitted under the SESP suffered major damage, and they were all able to be used as immediate community shelter, field hospitals and relief centres. Fortunately, the earthquake occurred on a Saturday when schools were not in session. The pioneering work of SESP and the demonstration of its effectiveness in the 2015 Gorkha earthquake has created the context for a national Comprehensive School Safety Framework, which is being adopted throughout the country.

7.3.5 Summary

There is some progress in the strengthening of school buildings in a number of countries. The OECD and World Bank Programmes, coupled with the efforts of national governments, have led to campaigns for the assessment of school buildings in some countries. However, in general, the extent of actual school retrofitting which has so far taken place is still very limited in relation to the need. And in most of cases, where retrofitting has taken place, it has yet to be demonstrated (by subsequent testing in an earthquake) that the techniques used have been successful in reducing the risk. This will not be clear for many decades.

7.4 Building for Safety Programmes

As we have seen in Chapter 3, in rural areas of the poorer earthquake-prone countries, the predominant form of wall construction is earth or stone masonry in a mud mortar. Some timber is normally used for the construction of roofs. These materials are locally available, and thus relatively affordable. But as commonly built, they have many weaknesses, including poor foundations, inadequate bonding of inner and outer wall layers, poor corner connections, degradation through water penetration, and poor connection between walls and roof structure (Coburn and Spence 2002). These together make them highly vulnerable to earthquakes, and the collapse of many such buildings was the main cause of the large death toll in many of the earthquakes of the last 20 years including the Mw7.6 2005 Kashmir earthquake in Pakistan, and the Mw7.8 2015 Gorkha Nepal earthquake in Nepal. NSET (2017) reports that the 2015 Nepal earthquake destroyed 490 000 houses, mostly traditional mud-brick or stone with mud mortar houses.

Eric Dudley (1988) has described a 1987 project to rebuild rammed-earth houses destroyed in an earthquake in the mountainous rural areas of Ecuador. The project involved training programmes for local builders, using essentially the existing traditional rammed earth technology, but with some modest modifications. These included the adaptation of the timber moulds used, to remove a corner weakness and the incorporation of timber ring-beams (Figure 7.3). Dudley reports, though, that as much as the technological improvements, it was the involvement of each community in the decision-making that enabled the project's success. In this case, the community leaders had responsibility for identifying the houses to be rebuilt, which master-builder from each community should attend the training courses, and the allocation of the master-builders' time. The project achieved a limited objective – the rebuilding of 1500 houses across 80 communities in 6 months – but Dudley (1988) argues that true development will only take place through the strengthening of indigenous institutions directly accountable to the local people.

Figure 7.3 Use of modified mould for rammed earth construction in Ecuador, 1987. *Source:* Eric Dudley. Reproduced with Permission.

Later, based on work over four years at the Centro Sinchaguasin in highland Ecuador, Dudley proposed, in his book *The Critical Villager* (Dudley 1993) three tests for judging proposed technical aid interventions in rural societies:

- Does it make sense? Is it *reasonable* in terms of the intended beneficiary's own rationale?
- What is it? Can the idea be *recognised* – does it have a name and are its limits clearly defined?
- Is it worthy of me? Is the idea *respectable* – is it something that 'people like us' do?

These principles (the three rs), says Dudley, do not describe a recipe for technical aid. But they emphasise the importance of judging proposed technical interventions from the perspective of the proposed beneficiaries, as well as their technical benefits.

The UN Sendai Framework for Disaster Reduction (UNISDR 2015) also recognises the important role of local community involvement and the incorporation of traditional knowledge in two of its guiding principles, as follows: 'it is necessary to empower local authorities and local communities to reduce disaster risk including through resources, incentives and decision-making responsibilities' (Clause 19f) and 'disaster risk reduction requires a multi-hazard approach and inclusive risk-informed decision-making based on the open exchange and dissemination of disaggregated data. . .complemented by traditional knowledge' (Clause 19g). And in another of its guiding principles, the Sendai Framework emphasises the importance of building back better. 'In the post-disaster recovery, rehabilitation and reconstruction phase it is critical to reduce disaster risk by Building Back Better, and increasing public education and awareness of disaster risk'.

All three of these guiding principles are exemplified in the successful work of NSET in Nepal before and after the 2015 earthquake sequence, and in the work of NCPDP in India, in the areas affected by the 2001 Gujarat earthquake, described below. However, in many cases, technical innovations in the aftermath of earthquakes have not been as useful.

In Indonesia, following the Mw9.1 Indian Ocean earthquake and tsunami of 2004, a large number of international aid agencies established their own programmes for rebuilding destroyed houses. These were reviewed three years later by Boen (2008), and although most were thought to be adequate to meet life safety criteria (Arup 2006), all involved the use of imported technology, with the result that they were quite unsuitable as models for others to construct their own houses (Figure 7.4). As Boen states it:

> It can be concluded that the alien construction methodologies are not sustainable because they are not culturally appropriate, irrespective of whether the as-built houses are earthquake resistant. Sustainability and culturally appropriateness shall be the main justification not to introduce such alien construction methodologies.

Boen concludes that the overall reconstruction programme has done little to reduce the risk from future earthquakes. The importation of such alien technologies after disasters has been a widespread, and much criticised, practice in the past (Davis 1978), and recent examples from Aceh, Haiti and Wenchuan suggest it is still prevalent today, largely because donor agencies (or national governments) still often think in terms of achieving production targets for the time-limited reconstruction programme rather than the longer-term benefit to the local communities.

Figure 7.4 'Alien technology' in reconstruction. Reconstructed houses in Aceh, Indonesia, using light steel frame. *Source:* Photo: T Boen.

7.4.1 NSET, Nepal

In Nepal, builder training has been an integral part of the work of NSET (Box 7.1) in its School Earthquake Safety Programme (see Section 7.3) since the start of SESP in 1998. The programme from the outset recognised the importance within local communities of the building craftsmen for both implementing and sustaining earthquake-resistant construction. The training has included both classroom and hands-on sessions, and course materials designed to address local issues have been prepared (Bothara et al. 2018). The 2015 earthquake sequence gave a big boost to this programme. NSET were commissioned by the Government of Nepal to provide technical assistance to a programme of rebuilding, called Baliyo Ghar (Safe house). The programme does not fund the reconstruction: it provides the technical assistance needed for owner-driven reconstruction. A five-year programme (2015–2020) was envisaged (NSET 2017) in which 50000–60000 masons would be trained for the construction and retrofitting of buildings. This huge programme is based on the following main principles:

- Build back better, requiring that houses reconstructed should have earthquake-resistant features as required in the National Building Code's guidelines (Government of Nepal 1994b).
- Technologies used should be based on the promotion of local construction materials and indigenous technologies.
- Given the scale of programme required, emphasis is on developing training materials and the training of trainers.
- Community-based public awareness programmes should always accompany the builder training programmes.

Box 7.1   Profile: Amod Dixit

Amod Dixit: earthquake technology in the service of the local community

Amod Dixit is the founder and Secretary-General of NSET, the National Society of Earthquake Technology in Nepal, which was founded in 1994. NSET is now known and respected throughout the world. In reply to the question 'how did NSET begin', Amod replied that there were two episodes which were the turning points for him and for NSET. In the first episode: 'I was a government geologist trying to map the geological effects of the August 1988 earthquake a month after the event, and I met a warehouse keeper in the town of Katari, who said to me "My son died, my wife is injured badly, and I feel insecure about my future: what is the use of your science to me? Why does the government employ you scientists to mock at our misery?" The keeper, a low-ranking staff, was looking after the earthquake-dilapidated warehouse building when all the higher-ranking staff went on emergency leave to take care of their families.'

'The second episode was a conversation with Brian Tucker of Geohazards International, which took place at a workshop in Bangkok in January 1993 organized by the World Seismic Safety Initiative (WSSI). I told him "I am tired of all expatriate experts teaching me what to do, when I know exactly what to do. What I need is your support in risk reduction work that I can't do in Nepal because of want of experience, know-how and resources." He replied, "*I understand you, Amod, and have decided to help you as you want. Let's work together starting next week.*"'

Amod writes of these two episodes, 'The first wrenched my heart and challenged me, a scientist with European education, with two questions, "what could I have done to help the simple warehouse-keeper to avoid death to his child and injury to his wife", and "what should I be doing henceforth". I felt helpless as I was unable to answer either question. The second episode provided me the much-needed compassionate intellectual energy to start doing something in earthquake disaster risk reduction.'

Amod's dream of earthquake risk management in Nepal started one week after his return from Bangkok when Brian Tucker arrived in Nepal with Richard Sharpe of BECA, New Zealand (then team leader of the building code development project) to support preparatory works towards establishment of a new organization by a group of Nepalese engineers and geologists, NSET. NSET received its legal status with the government as a civil-society organization (NGO) on 18 June 1994. Its prime motivation was to assist the government and the people to improve seismic performance of Nepalese buildings, including the non-engineered structures made in bricks, stone and timber in mud mortar.

NSET started its first formal project – the Kathmandu Valley Earthquake Risk Management Project (KVERMP) in September 1997 with financial support from the US office of Foreign Disaster Assistance (OFDA). KVERMP included a simplified earthquake risk estimation based upon information on earthquake hazard, and its use for a wider earthquake awareness; the development of a ten-point action plan for earthquake risk management for Kathmandu Valley, and the start of a locally driven programme of seismic retrofitting of school buildings.

The success of KVERMP provided opportunities to start several innovations that have now become incorporated into the government's annual budget. Seismic retrofitting of school buildings has developed into a School Earthquake Safety Program (SESP) (Figure 7.2). NSET also organises a mason training programme for earthquake-resistant construction and seismic retrofitting (Figure 7.5). Earthquake Safety Day (ESD) programmes are observed annually by the government with participation by all. On Earthquake Safety Day, shake table demonstrations, an annual symposium on disaster risk management and an exhibition on earthquake risk reduction and preparedness are organised in all district headquarters. Mobile earthquake clinics and earthquake orientation classes as well as 'earthquake walks' are organised throughout the year at various places, and print as well as electronic media and radio stations air public service announcements, videos and talk programmes sponsored by the government, private sector businesses including hoteliers and travel and tour agencies.

The legacy of KVERMP continues – NSET has been continuously supported financially in the past two decades by the government, international development partners and also by universities and research organisations in many countries. NSET was invited to work with government and other agencies in Gujarat, Bam, Banda Aceh and Pakistan for earthquake reconstruction using owner-driven approaches. Through these collaborations, NSET has learned much which has helped to provide services to Nepal in helping people in disaster preparedness, both in 'building back better' following the 2015 Gorkha earthquake and also in 'building better now' by assisting municipalities in building code implementation.

NSET's success derives from its focus and community-driven approaches, integrating understanding of global developments in earthquake science and technology with indigenous knowledge for disaster risk management.

By March 2017, 131 mason training courses had been carried out, training over 3000 masons (Figure 7.5). Many of these masons were not previously employed in building, and a significant proportion of them were women.

In the course of this programme, the training moved into hilly areas where the local technology differs from those covered in the national code guidelines, and for these areas revised techniques have been developed, which include the use of two-storey buildings with a timber upper floor. There have been other problems, too. In the early years of implementation, community workers found people sceptical that earthquake-resistant technologies could be achieved without massive investment. In some areas, the demand for materials caused an increase in price and the need for timber-generated conflict with forest conservation officials.

Figure 7.5 Graduate of the NSET mason training programme applies lessons learnt. *Source:* Photo: NSET. Reproduced with Permission.

7.4.2 NCPDP, Gujarat, India

Another good example of builder-training for earthquake-hazard mitigation is the work of the National Centre for Peoples Action in Disaster-Preparedness (NCPDP) based in Ahmedabad, India (http://ncpdpindia.org). Founded and motivated by an architect/engineer couple Rajendra and Rupal Desai (see Box 7.2), NCPDP has been organising builder training programmes since 2006, based on experience of such training programmes acquired through projects following the Mw6.2 1993 Latur earthquake, the Mw7.7 2001 Gujarat earthquake and the Mw7.6 2005 Kashmir earthquake. At the heart of the project is the experience of observing the reconstruction programme following the 1993 Latur earthquake, as a part of a government disaster-assessment team. The Latur earthquake, the first in the area for many centuries, destroyed dozens of villages built with the traditional materials of the area, stone, mud and timber. The disaster was blamed, both by officials, engineers and the survivors themselves on the use of these local materials, and the government policy on reconstruction became one of providing estates of houses built with brick and reinforced concrete walls with concrete or metal sheet roofs. These 'suburban' style houses were alien to the living culture of the local people, which is based on a courtyard house around a veranda, providing shade and privacy, cool in winter and summer. The new houses were tiny and expensive; the owners had no involvement in planning and designing; privacy and security were lost; the houses were hot in summer and cold in winter; and all the materials were bought on the market and transported to the site.

NCPDP believed that an alternative approach to rebuilding was possible which was both more culturally sensitive and environmentally sustainable while also achieving adequate earthquake resistance. They stayed in Latur district for several years and developed, in different villages, ways to build not only using essentially the local technologies and materials but also providing far better earthquake-resistance. For instance they proposed

---

Box 7.2   Profile: Rajendra and Rupal Desai

Rajendra and Rupal Desai: disaster-resistant rural housing in India

Early in their professional lives, Rupal, an architect and Rajendra, a structural engineer, were working on modern industrial construction projects in the United States. But since the 1970s, they became increasingly concerned about environmental sustainability. This led them, in 1984, to make the momentous decision to return to their native India and put their technical background to use for 'the betterment of the people and community in rural India' and at the same time to stop working for monetary gain.

In the early years, they worked on tribal housing in their home state of Gujarat, learning about the local economy and looking for housing options which were simple to learn and execute, contributed to the local economy, and had a low embodied energy. Their ideas were influenced by the principles of Gandhian economics.

Their lives were changed again when, in 1993, they were asked to join a post-earthquake damage assessment mission following the devastating Latur earthquake in Maharashtra. This led them to spend the next six years helping people in the affected villages to better protect themselves from future earthquakes, by improving the vernacular house building technology based on stone, earth and timber. They worked on the training of over 1000 local artisans and 550 government engineers, and retrofitted 150 village houses. And they conducted local education and awareness programmes through street plays, exhibitions and rallies, shock table demonstrations and a variety of publications, to rebuild people's confidence in local building technologies.

Through setting up the National Centre for Peoples Action in Disaster Preparedness (NCPDP) in 2000, they were able to carry out similar activities at a larger scale after the 2001 Gujarat earthquake, training over 8000 artisans, and 1400 government engineers, capacity building in 478 severely affected villages, rebuilding 4 entire villages and retrofitting over 400 public buildings including schools.

In the years following, they worked on the development of simple improvements to existing rural building technologies in the Himalayan regions of Uttarakhand, Kashmir

(Continued)

Box 7.2 (Continued)

and Nepal, combining work on specific buildings with the training of local artisans and engineers and producing guidance documents.

Returning to Gujarat in 2004, under their guidance, NCPDP has been conducting short building skill up-gradation programmes for building artisans in the rural areas, for practising masons and bar-benders (Figure 7.6). By 2019, these courses had been attended by over 4500 artisans. Manuals and guidance documents for repair and retrofitting of houses in many parts of India have been prepared, in several languages. Their work has been instrumental in establishing a Building Artisan Certification Scheme for India's Ministry of Housing.

During these years, Rupal and Rajendra have worked closely together as a team, aiming to achieve certain principles:

- Use of ecologically sound and culturally suitable appropriate technology with minimal carbon footprints for rural housing.
- Use of local materials, local artisans and improved local vernacular technology.

Both are involved in the training programmes. Rupal focuses on documentation and developing building designs, while Rajendra works on conceptualisation, technology development, policy interventions and dealing with national and international agencies.

Rupal and Rajendra as they put it, have led 'a path of search and research. We firmly believe that only building local capacity in the use of traditional building technology with the infusion of modern science and technology in line with Gandhian principles of simplicity and self-reliance can bring longer-lasting affordable disaster-resistant houses. This would help restore people's pride in them and spread the safety net with minimal contribution to global warming to ensure sustainability for future generations'.

Figure 7.6 Artisan skill up-gradation programme in Gujarat, India, 2019. *Source:* Photo: NCPDP.

using reinforced concrete lintel bands at the tops of walls and adding bracing elements to timber frames. They also developed training courses for masons, and importantly, techniques for demonstrating to house-owners the effectiveness of the improved techniques against earthquake shocks. And, importantly, they were able to recruit community leaders to adopt their technologies and give them the kind of respectability suggested by Eric Dudley, creating a demand for the new techniques to replace the initial scepticism (Desai and Desai 2010).

Experiences in rebuilding after the 2001 Gujarat earthquake and the 2005 Kashmir earthquake showed that this approach (though with locally adapted techniques) could be widely adopted in the earthquake-prone rural areas of India. Since 2006, when NCPDP was founded, artisan skill 'up-gradation' programmes have been carried out in many villages in Gujarat beyond the areas affected by the 2001 earthquake. The three-day programme, which takes place in or very near the village where the artisans live, focuses on the issues most critical for quality improvement and disaster-resilience in the traditional building technology of the area. The first day is devoted to classroom lectures supported by posters, videos and working models; the second day to hands-on sessions, and the third day to recapitulation and oral and written tests prior to the award of a certificate. In 15 years since the programme started, more than 5000 artisans have participated and been awarded a certificate, and the programme continues (NCPDP 2019) (Figure 7.6).

However, mounting these programmes has problems. NCPDP is an NGO, and receives all its funding from private and corporate donors. The course needs to take place over just three days, because even though participants are paid a small daily stipend of 200 rupees, this is a fraction of what they are typically paid for a day's work, and they cannot afford to stay longer. A social mobiliser is used to recruit artisans to participate.

Nevertheless, the work of NCPDP has had an influence on the design of similar builder training programmes in other parts of India and neighbouring countries including Nepal. Such programmes need to be much more widely replicated.

7.4.3 Summary

Although there is a general recognition of the need for builder training programmes, both at the UN and World Bank level, and by many earthquake specialists worldwide (Jain 2016), they are not yet very common, for a number of reasons. First, the training programmes do depend on sophisticated skills at a technological and architectural level: skills which are not taught today in schools of architecture and engineering in developing countries, where the focus is on building for the urban sector in modern and manufactured materials. They also depend on the development of skills to communicate technical details to builders and house-owners with little formal education. This also creates a demand for new types of building guidance documents designed to reach rural communities. Many such guidance documents have been developed to support builder-training programmes, some of which have not been adequately tested and fail to communicate the intended messages (Dudley and Haaland 1993).

In addition, achieving an adequate level of earthquake-resistance using traditional technologies, while technically feasible, cannot be entirely cost free. Programmes have involved the addition of timber ring-beams, or small quantities of steel reinforcing bars or gabionbands and other modifications to otherwise traditionally built structures. In the absence of subsidised programmes (to date mainly adopted only in post-earthquake reconstruction), the owners must be persuaded that the costs of such extras is justified by the extra safety they provide from a hazard they are likely to perceive as distant. And, as seen in Nepal, extensive use of local materials, especially timber, may turn out to be in conflict with local conservation policies. However, such programmes are urgently needed: and a major increase in the funding of such programmes, both nationally and internationally, is now needed to make an impact on the level of earthquake risk at a national level. Appreciation of the value of maintaining the vernacular way of building, coupled with an awareness of the danger of existing ways of construction, is needed to create the context for application of better building techniques. How to generate such public awareness is a problem common to building-owners worldwide and is discussed in the Section 7.5.

7.5 Public Awareness of Earthquake Risk: Creating a Safety Culture

A strong public awareness of earthquake risk, leading to what may be called a 'safety culture', is an essential starting point for all of the activities for improving the earthquake safety of buildings discussed above. Action on implementing earthquake codes, strengthening existing buildings and adopting earthquake-resistant techniques in the construction of non-engineered buildings all depend on a public awareness of the earthquake risk. Prior to an earthquake, the public need to know about actions they can take to make their home, community or place of work safer in the event of an earthquake; if an earthquake occurs, the public need to be informed and trained in the actions they can take to protect themselves; and they need to know what they can do in the aftermath of an earthquake to help their community in the relief and recovery stage.

Earthquakes are rare events. In any one location, even in the areas most prone to earthquakes, a level of ground shaking able to cause serious building damage is unlikely to occur more than once or twice in a lifetime. For other hazards, such as fires and windstorms, which occur more frequently, safety-consciousness is partly learnt through experience. And, unlike other hazards which usually give some warning, earthquakes generally occur without warning signals. Thus, maintaining public awareness is a matter for collective community action.

It is often said that, following a damaging earthquake, there is a 'window of opportunity' of just a few years to take action leading to better safety in the future. California's Field Act requiring the preferential earthquake strengthening of schools was passed within a few months of the occurrence of the Mw6.4 1933 Long Beach earthquake, even though it had been proposed a decade earlier (Alesch and Petak 1986). And, as reported earlier, school strengthening in Japan was given a big boost by the occurrence of the 1995 Kobe earthquake. In the immediate post-event years, public awareness of the risk is high. But after some years of little or no earthquake activity, as has been the case in California in the years since the 1994 Northridge earthquake, public concern tends to decline, and other more immediate concerns – such as wildfires in California and Australia, and windstorms in many tropical countries – tend to occupy public and media attention. This is particularly the case as these hazards are expected to grow with global warming. But from what we now know about earthquake occurrence, it is clear that, on active faults, the risk of a new event also grows with the passage of time since the last one, because of the accumulation of stress in the earth's crust (see Chapter 4). At such times, public awareness campaigns become the essential means to maintain public alertness to the danger of the next occurrence of 'the big one'.

Public awareness does not of course necessarily lead to action. In their book Communicating Building for Safety, Eric Dudley and Ane Haaland (1993), have suggested that there is a universal process of adopting new ideas, in which awareness has to be followed by interest, and then trial and evaluation, before adoption can occur. They suggest that 'why-to' information is more important in the first two stages, and that 'how-to' information becomes important only at the trial and evaluation stages. Different channels of communication are also needed at each stage.

Public awareness programmes thus take a great variety of forms, depending on the existing level of awareness, on the economic situation of the public addressed, the available channels of communication, and the age of the intended participants. Schoolchildren are especially important, as they represent the future of society, and earthquake awareness can be incorporated into the regular teaching curriculum. School-teaching about the tsunami risk enabled the appropriate evacuation response of many people in both the 2004 Indian Ocean and the 2011 Tohoku events (see Box 4.1). But public awareness programmes also need to be designed to reach homeowners, businesses and community organisations. Some examples of public awareness programmes, from California, Japan, Europe and Nepal are given in the following sections.

7.5.1   California: The Great Shakeout

The Great California Shakeout (Jones and Benthien 2011), (http://www.shakeout.org) is a large-scale public awareness event which takes place each year. Originally, focussed on Southern California, it now includes participants in every region of the United States, Canada, Japan, Italy, New Zealand and other countries. The first Shakeout event in 2008 included over 5 million participants taking part in simultaneous earthquake drills in schools and universities, homes and businesses, and the 2019 event included as many as 67.5 million people around the world, 10 million of whom were in California. The third Thursday in October is the day most shakeout drills take place. They include practicing Drop, Cover and Hold On, the earthquake response action recommended for California where injury from collapsing ceilings and furniture is far more likely than complete building collapse. But participant organisations are also encouraged to carry out actions to make their space safe, by investigating non-structural hazards, securing furniture, storing water and having fire extinguishers, renewing insurance policies, as well as checking the structural adequacy of their buildings. And emergency response teams use the same day to practice their own drills.

The Shakeout actions are promoted and supported by a variety of consistent sources of information. An important element of the event has been the development and publication of an earthquake scenario based on the consequences of a Mw7.8 earthquake on the southern part of the San Andreas Fault, through which the anticipated damage to buildings and infrastructure were estimated and mapped, and the consequences in the short and longer term were detailed. The scenario estimated a loss of about $213bn, and about 1800 fatalities.

More than 300 scientists collaborated on developing this scenario. The aim was to get people talking about the impact and discuss how it could be reduced (Jones and Benthien 2011), Box 9.1.

Messages were designed to be communicated by a set of visual messages, using newspapers, broadcast and social media. The Shakeout website is the centre of the recruitment effort, and through it detailed data is made available in a form appropriate to different types of participating organisations, in different parts of the country and in other countries, and in a number of languages (Figure 7.7).

The aims the original 2008 Great California Shakeout were to

- Register five million people to participate in the drill – one quarter of the population of Southern California.
- Change the culture of earthquake preparedness in Southern California – convince people that much of the loss estimated by the scenario was preventable.
- Reduce earthquake losses – the ultimate goal of the first two aims, though not easily measurable.

However, the extraordinary success of this enterprise in attracting participants for its now annual drills suggests it is seriously contributing to generating the necessary safety culture throughout the United States and beyond.

7.5.2 Japan: Disaster Prevention Day

Japan faces the threat of several different types of disaster, windstorm and fire as well as earthquakes and the accompanying tsunamis. Since 1960, September 1, the day of the occurrence of the massive Mw8.1 Great Kanto earthquake of 1923, has been Disaster Prevention Day. In 2014, about 2.3 million people took part in drills throughout Japan.

The Tokyo Metropolitan government has compiled and distributed freely to citizens and visitors, a manual called Disaster Prevention Tokyo also translated into English (Tokyo Metropolitan Government 2020) Figure 7.8. Its cover page warns citizens that there is a 70% possibility of an earthquake directly hitting Tokyo within the next 30 years and asks

Figure 7.7 Drop, Cover and Hold On graphic from the Great California Shakeout website (http://www.shakeout.org). *Source:* DROP, COVER, AND HOLD ON!, PROTECT YOURSELF. SPREAD THE WORD. www.shakeout.org. © Southern California Earthquake Center.

'Are you prepared?'. The manual contains easy to understand information on how to prepare for and respond to a disaster.

Separate sections of the manual contain illustrated guidance on stockpiling, preparing inside the home (including advice on securing and stabilising furniture, seismic retrofitting and fire prevention) and on understanding the neighbourhood and community networking.

7.5.3 Europe: Developing and Assessing Earthquake Awareness Dissemination Programmes

Public awareness of earthquake risks is not as high in Europe as it is in Japan, but the level of awareness varies considerably depending largely on recent earthquake experience. Two recent EU-funded projects have brought together research institutes in Italy (INGV), Portugal (LNEC) and Iceland (EERC) to develop earthquake-awareness and mitigation strategies, apply them through pilot training projects in schools, and to assess the level of awareness before and after the training and the level of application of the mitigation strategies. The UPStrat-MAFA Project (2012–2014) was concerned with the development of strategies for urban disaster prevention in relation to structural damage and economic impact (Musacchio et al. 2016), while KnowRISK (2016–2017), was focussed on non-structural damage. KnowRISK targeted areas where major structural damage was unlikely, but non-structural damage (architectural elements – cladding, ceilings, electrical and mechanical systems, and building contents) was likely to be the major cause of damage. In both

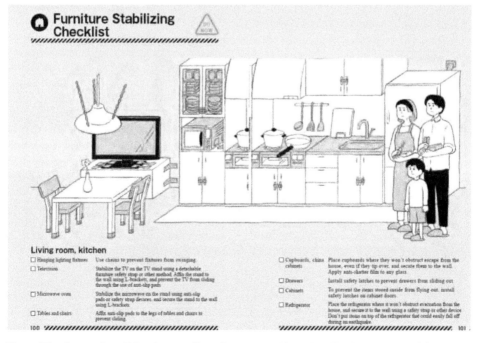

Figure 7.8 Extract from Tokyo Metropolitan Government Manual on Earthquake Preparedness, giving advice on stabilizing furniture in the home. *Source:* Tokyo Metropolitan Government (2020).

projects, pilot projects took place in schools, but a variety of guidance documents, events and demonstrations were available for the general public.

In Portugal, lack of any significant earthquakes in the recent past means that earthquake awareness is low; it is higher in some parts of Italy where there have been recent earthquakes, but still low in others. In the Southern Iceland Seismic Zone, three recent damaging earthquakes have resulted in a higher level of earthquake awareness. In the Mw6.6 South Iceland earthquake of 2000, injury was reduced because at the time of the earthquake, many of the residents were outside their homes (Platt et al. 2019). In Italy, the one country with recent survey data, a national sample of 4000 people showed that only 17% of people living in lower-hazard area, and surprisingly, only 6% of people living in higher hazard areas have what was considered adequate perception of the risk (Crescimbene et al. 2016).

7.5.4 Nepal: Awareness Raising

Although one of the poorest of earthquake-prone countries, significant efforts have been made in Nepal in recent years (beginning well before the 2015 earthquakes) to raise public awareness of earthquake risk. The Government of Nepal believes in a bottom-up approach to achieve effective disaster risk reduction, and January 15 (the day of the devastating Mw8 earthquake in 1934) has been declared as Earthquake Safety Day. The government, supported by NSET and local authorities, has carried out a variety of public awareness events, including publications and distribution of brochures with, street theatre events, shock table demonstrations (Figure 7.9), and dissemination of press information (Bothara et al. 2018).

Some impacts from this public awareness activity have been (NSET 2017):

- A significant increase in the awareness of earthquake risks in surveyed municipalities
- Engagement of many people in non-structural mitigation activities
- Increasing observed success in urban areas in the level of building code implementation

7.5.4.1 Summary

The existence and success of public awareness campaigns is not necessarily a satisfactory measure of a growing safety culture, nor necessarily results in implementation of risk reduction measures. Indeed, measuring the effectiveness of public awareness efforts is not easy. A study carried out in 2007 by University College London in three earthquake-prone countries, USA, Japan and Turkey found, in each country, a poor correlation between earthquake risk perception and the adoption of a range of seismic adjustments (Rossetto et al. 2011). In Nepal, a Knowledge, Attitude and Practice (KAP) survey conducted in 10 municipalities in 2011, found that while levels of knowledge and positive attitude to earthquake risk mitigation were between 60 and 80%, the proportion taking any resulting action was relatively low, generally around 20–30%.

This is not to say that earthquake-awareness campaigns and drills are ineffective. In Mexico, drills take place in schools, offices hotels and private homes on September 19 each year – the anniversary of the devastating earthquake of 1985. These along with stricter and better-enforced building codes in urban areas, and the promotion of confined masonry construction for non-engineered buildings, have led to considerable improvement in

Figure 7.9 Shock-table demonstration to convince people of the effectiveness of earthquake-resistant construction in Nepal. *Source:* Photo: NSET. Reproduced with Permission of NSET

earthquake-preparedness and risk reduction. By coincidence, the 2017 Mw7.1 Puebla earthquake, with its epicentre 120 km from Mexico City, occurred on September 19, the same day as the earthquake drills, which had taken place two hours earlier. Though 370 people were killed, the earthquake provided evidence that Mexico has achieved much in creating a safety culture since the disaster of 1985 (Lahiri 2017).

The extent of action taken to improve the safety of buildings differs widely country by country and regionally within countries. It seems to depend primarily on

- The experience and impact of recent earthquakes in that region
- The availability of resources to pay for the action needed – at a household, business or municipal scale
- The extent of public concern, and resulting willingness to devote available resources to earthquake protection rather than other types of improvements.

In developing countries, where resources are scarce, financial support by development agencies or NGOs is likely to be needed to translate awareness and concern into effective mitigation programmes.

References

Alesch, D. and Petak, W. (1986). *The Politics and Economics of Earthquake Hazard Mitigation: Unreinforced Masonry Buildings in Southern California*. University of Colorado.

Arnold, C. (2012). *Timber Construction*. World Housing Encyclopedia.

Arup (2006). *Aceh & Nias Post Tsunami Reconstruction – Review of Aceh Housing Program*. London: Ove Arup and Partners.

Boen, T. (2008). Reconstruction of houses in Aceh, three years after the December 26 2004 Tsunami. *Presented at the International Conference on Earthquake Engineering and Disaster Mitigation* 2008, Bandung, Indonesia.

Bogazici University, Istanbul Technical University, Middle East Technical University, and Yildiz Technical University (2004). Earthquake Master Plan for Istanbul.

Booth, E. (2014). *Earthquake Design Practice for Buildings*, 2e. London: ICE Publishing, Institution of Civil Engineers.

Booth, E. (2018). Dealing with earthquakes: the practice of seismic engineering "as if people mattered". *Bulletin of Earthquake Engineering* 16: 1661–1724.

Bothara, J., Ingham, J., and Dizhur, D. (2018). Earthquake risk reduction efforts in Nepal. In: *Integrating Disaster Science and Management* (eds. P. Samui, D. Kim and C. Ghosh). Elsevier.

BSSC (1997). *NEHRP Recommendations for Seismic Regulations for Buildings and other Structures*. Building Seismic Safety Council Report FEMA 302/303.

California Seismic Safety Commission (2004). *Seismic Safety in California's Schools: Findings and Recommendations on Seismic Safety Policies and Requirements for Schools*. Sacramento, California: California Seismic Safety Commission.

Cardona, O.-D., Alexander, D., Bhatia, S. et al. (2015). Seismic safety in schools: case studies. In: *Encyclopedia of Earthquake Engineering*. Springer-Verlag Berlin Heidelberg https://doi.org/10.1007/978-3-642-36197-5_406-1.

CEN (2004). Eurocode 8: design of structures for earthquake resistance. EN 1998.

CEN (2020). Eurocode 8: design of structures for earthquake resistance. EN 1998: Working Draft of Revised Part 1–1 and 1–2 TC250/SC8.

Charleson, A. (2008). *Seismic Design for Architects: Outwitting the Quake*. Oxford and New York: Taylor and Francis.

Coburn, A. and Spence, R. (2002). *Earthquake Protection*. Wiley.

Crescimbene, M., La Longa, F., Peruzza, L., Pessina, V., Pino, N., 2016. The seismic risk perception in Italy compared to some hazard, exposure and vulnerability indicators. *Presented at the International Conference on Urban Risks* 2016, Lisbon, Portugal.

Davis, I. (1978). *Shelter After Disaster*. Oxford: Oxford Polytechnic Press.

D'Ayala, D., Spence, R., Oliveira, C., and Pomonis, A. (1997). Earthquake loss estimation for Europe's historic town centres. *Earthquake Spectra* 13: 773–793.

Deppe, K. (1988). The Whittier narrows, California earthquake of October 1, 1987—evaluation of strengthened and unstrengthened unreinforced masonry in Los Angeles City. *Earthquake Spectra* 4: 157–180.

Desai, R. and Desai, R. (2010). *Strengthening Peoples' Architecture Against Disasters*. NCPDP.

Dudley, E. (1988). Disaster mitigation: strong houses or strong institutions. *Disasters* 12: 111–121.

Dudley, E. (1993). *The Critical Villager: Beyond Community Participation*. London and New York: Routledge.

Dudley, E. and Haaland, A. (1993). *Communicating Building for Safety: Guidelines for Communixcating Technical Information to Local Builders and Householders*. London: Intermediate Technology Publications.

Fajfar, P. (2018). Analysis in seismic provisions for buildings: past, present and future. *Bulletin of Earthquake Engineering* 16: 2567–2608.

Government of Nepal (1994a). Nepal national building code: NBC 000:1994 requirements for state of the art design, an introduction. Minsistry of Physical Planning and Works, Government of Nepal.

Government of Nepal (1994b). Guidelines for earthquake-resistant construction: low strength masonry: nepal building code NBC 203:1994. Minsistry of Physical Planning and Works, Government of Nepal.

Hamburger, R. (2003). Building code provisions for seismic resistance. In: *Earthquake Egineering Handbook* (eds. W.F. Chen and C. Scawthorn), 11.1–11.28. CRC Press.

IPKB, İPCU (2020). ISMEP project update. IPCU. http://www.ipkb.gov.tr/en/what-is-ismep/b-component (accessed October 2020).

ISMEP (2014). *Retrofitting and Reconstruction Works*. Istanbul Project Coordination Unit.

Jain, S.K. (2016). Earthquake safety in India: achievements, challenges and opportunities. *Bulletin of Earthquake Engineering* 14: 1337–1436.

Jones, L. and Benthien, M. (2011). Preparing for a "big one": the great Southern California shakeout. *Earthquake Spectra* 27: 575–595.

Lahiri, T. (2017). Mexico held an earthquake drill two hours before its latest deadly quake hit. https://qz.com/1082239/mexico-held-an-earthquake-drill-two-hours-before-its-latest-deadly-quake-hit (accessed October 2020).

Moullier, T. and Krimgold, F. (2015). *Building Regulation for Resilience: Managing Risks for Safer Cities*. Washington, DC: GFDRR, The World Bank.

Musacchio, G., Falsaperla, S., Sansivera, F. et al. (2016). Dissemination strategies to instil a culture of safety on earthquake hazard and risk. *Bulletin of Earthquake Engineering* 14: 2087–2103.

NCPDP (2019). Annual Report on Building Artisan Skill Up-gradation Program.

NSET (2017). *Safer Society: NSET Report 2017*. Nepal: National Society for Earthquake Technology.

OECD (2004). Part V in *Keeping Schools Safe in Earthquakes* (eds. R. Yelland and B. Tucker). Paris: Programme for Educational Buildings, OECD.

OECD (2019). Housing stock and construction. HM1.1. OECD Affordable Housing Database – http://oe.cd/ahd, OECD - Social Policy Division - Directorate of Employment, Labour and Social Affairs. www.oecd.org/els/family/HM1-1-Housing-stock-and-construction.pdf.

Parsons, T., Toda, S., Stein, R.S. et al. (2000). Heightened odds of large earthquakes near Istanbul: an interaction-based probability calculation. *Science* 288: 661–665.

Petal, M., Wisner, B, Kelman, I. et al. (2015). School seismic safety and risk mitigation. In: *Encyclopedia of Earthquake Engineering*. Springer-Verlag Berlin Heidelberg https://doi.org/10.1007/978-3-642-36197-5_406-1.

Platt, S., Musacchio, G., Crescimbene, M. et al. (2019). Development of a common (European) tool to assess earthquake risk communication, chapter 37. In: *Proceedings of the International Conference on Earthquake Engineering and Structural Dynamics* (eds. R. Rupakhety et al.). Springer.

Porter, K.A. (2019). Society can afford seismically resilient buildings: a working paper for the Multihazard Mitigation Council of the National Institute of Building Sciences.

Rossetto, T., Joffe, H., and Solberg, C. (2011). A different view on human vulnerability to earthquakes: lessons from risk perception studies. In: *Human Casualties in Earthquakes: Progress in Modelling and Mitigation* (eds. R. Spence, E. So and C. Scawthorn), 291–304. Springer.

Smyth, A.W., Altay, G., Deodatis, M. et al. (2004). Probabilistic benefit-cost analysis for earthquake damage mitigation: evaluation measures for apartment houses in Turkey. *Earthquake Spectra* 20: 171–204.

Spence, R. (2004). Strengthening school buildings to resist earthquakes: progress in European countries. In: *Keeping Schools Safe in Earthquakes* (eds. R. Yelland and B. Tucker), 217–228. Paris: OECD.

Spence, R., Oliveira, C., D'Ayala, D., Papa, F., and Zuccaro, G. (2000). The performance of strengthened Masonry buildings in recent European earthquakes (paper 1366). *Presented at the 12th World Conference on Earthquake Engineering* 2000, Auckland.

Tokyo Metropolitan Government (2020). Lets get prepared: disaster preparedness actions. https://www.metro.tokyo.lg.jp/english/guide/bosai/index.html (accessed 2020).

UNISDR (2015). Sendai framework for disaster risk reduction 2015–2030. www.undrr.org/publication/sendai-framework-disaster-risk-reduction-2015-2030 (accessed July 2020).

Wisner, B., Kelman, I., Monk, T. et al. (2004). School seismic safety: falling between the cracks. In: *Earthquakes* (eds. C. Rodrigue and E. Rovai). London: Routledge.

World Bank (2016). *Making Schools Resilient at Scale: The Case of Japan 2016*. Washington: GFDRR, World Bank.

World Bank (2017). Global program for safer schools. http://www.worldbank.org/en/topic/disasterriskmanagement/breif/global-program-for-safer-schools.print (accessed July 2020).

8

Successes and Failures in Earthquake Protection: A Country Comparison

8.1   Introduction: The survey

In this chapter, we will review the achievements in earthquake protection, country by country, for many of those countries at greatest risk from earthquakes. The approach to obtaining the information was to invite one or more highly experienced professionals in each country to respond to a simple survey. A previous survey along these lines was conducted in 2007 (Spence 2007). For many countries, therefore, the current survey, carried out for this book, was an update of the 2007 survey, with information provided by the same respondent. However, the current survey includes several countries which were not included in 2007, and in some cases, the response was from a different expert.

The questions were deliberately kept short and general, to enable respondents to deal with them in the way best suited for them. The questions were as follows:

1) In your view what are the most significant success and failures of earthquake protection in your country in the last 10 years?
2) Have there been changes in the extent of implementation of earthquake-resistance aspects of the codes of practice for the design and construction of new buildings since 2007?
3) What do you estimate is now the proportion of unsafe buildings (pre-code buildings, or those built without applying codes) in the current urban and rural building stock?
4) Have there been, since 2007, new or extended programmes to assess and upgrade unsafe buildings (either public buildings, e.g. schools and hospitals, or residential buildings), and how successful have these programmes been?

The invitations to contribute to the survey were sent in May 2019, and responses were received during summer 2019 from 39 respondents in 28 different countries. The countries were chosen because of their experience of damaging earthquakes in the last 20 years. Many of the respondents were leading academics, but a few were practising professional engineers. This was a very encouraging response, and we are deeply indebted to all the respondents for their contributions, acknowledged at the end of this chapter.

Reflecting on these responses, it seems appropriate to separate the countries into three separate groups. The first group we will refer to as the 'High Achievers'. These countries have made demonstrable progress in tackling their earthquake risk, and reducing it to a much lower level than existed 50 years ago, and have continued to make progress in the last 10 years. Japan, the US State of California, New Zealand and Chile belong to this group.

A second group of countries has made some progress, but it is slow, and limited, focussed on earthquake science and hazard mapping, improving earthquake codes and their enforcement in new buildings, and in the training of engineers. Overall risks are lower than they were, but many high risks remain, and much remains to be done to raise the awareness of the public, and to direct public and private resources to upgrading the existing building stock. In this category of 'Limited Achievers' are most of the European countries, and the group also includes China, Turkey and Mexico.

The third group of countries are those where, in spite of campaigning and action by a group of dedicated professionals, not enough is being done at a national level or by individuals to control risks, which continue to rise. In this 'Continuing and growing risks' category are Colombia, Indonesia, India, Nepal and several other countries.

The allocation of the countries to each of the three groups was done on the basis of how well the country response matches the description of the group given above. The following sections attempt to summarise, country by country, what has and has not been achieved, according to the reports and opinions of our survey respondents. In discussion of each country, where the level of the earthquake hazard is referred to, it is measured by the peak ground acceleration (PGA) with a return period of 475 years (or 10% probability of exceedance within 50 years). PGA is discussed in greater detail in Chapter 4. Definitions differ, but in this book (following EC8), regions with an expected 475-year PGA exceeding $4\,m/s^2$ are classified as having 'very high seismicity', those with between 2 and $4\,m/s^2$ have 'high seismicity', those with between 1 and $2\,m/s^2$ have 'moderate seismicity', and those with a 475-year PGA less than $1\,m/s^2$ have a 'low' or 'very low' seismicity.

Where the magnitude of an earthquake is stated, this is expressed, where possible, in terms of moment magnitude (Mw), now the most commonly used measure, as explained in Chapter 4.

8.2 High Achievers

8.2.1 California

California is justifiably seen as one of the global leaders in developing and implementing seismic safety policy. It provides a good example of how the experience of numerous major damaging earthquakes, since the 1906 San Francisco event (estimated as M7.9), have acted as catalysts to state and city legislation which has led to subsequent action. The earthquakes have also stimulated research programmes leading to improved methods and codes of practice for new buildings. Each successive earthquake has not only provided a test for new methods but also demonstrated continuing weaknesses, leading to further research. Since the Mw6.9 1989 Loma Prieta earthquake, there was a high level of public awareness and expectation, but (until July 2019) there has been no earthquake with a magnitude greater than Mw7, providing an opportunity for cities to consolidate their preparedness planning. California's wealth is a great asset to enable ideas to be turned into action in upgrading the existing building stock. But democracy and local autonomy have often been obstacles to state-wide action. Thus, even in California, our respondents identified some failures alongside the successes.

Successes identified in the earlier (2007) survey were of three types: legislation leading to action; research and development in methods of building; and market-driven mitigation. In each of these areas, there has been further progress in the last decade. School-, university- and hospital-strengthening programmes have made substantial progress, retrofits have been completed, and older buildings continue to be re-evaluated. New legislation is in place in Los Angeles requiring the strengthening of 'soft-storey' apartment buildings, and older reinforced concrete buildings – though with several decades to complete the work (Mayor of Los Angeles 2016), (Figure 8.1). Most of the older (pre-code) masonry buildings in both of the largest cities, San Francisco and Los Angeles, have been strengthened or replaced. And state-wide, through licensing of the professionals involved, application of Codes of Practice and building regulations in design and implementation on site is at a high level.

High-level research has continued, both on the characterisation of ground motion and on the performance of buildings, leading to further development of the codes and associated guidance documents. And the Earthquake Engineering Research Institute continues, especially through its Learning from Earthquakes programme, to ensure that the observations from damaging earthquakes everywhere in the world are investigated and documented for the benefit of the global engineering profession.

The Great California Shakeout (Jones and Benthien 2011) (https://www.shakeout.org/california) is a now annual state-wide emergency preparedness event providing communities, companies and individuals an opportunity to test their readiness for a big earthquake. Shakeout now has massive participation: 10 million Californians participated in 2019 and more than 60 million worldwide. It is described in Chapter 7.

The development of a resilience concept of earthquake safety is seen by several respondents as a very important step forward. The idea is that rather than dealing with earthquake

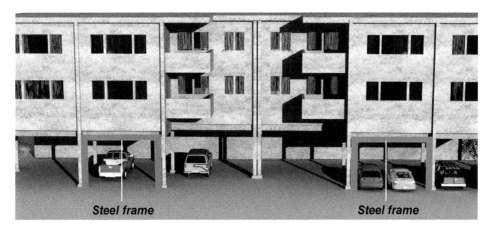

Figure 8.1 Soft storey timber apartment block in San Francisco, retrofitted with steel frames, 2020. *Source:* Photo: WCPC Inc, San Francisco.

safety on a building-by-building basis, earthquake risk and preparedness should be viewed on a community-wide scale. One important aspect of this is the recognition that current codes, while providing for life-safety, can cause key buildings and services to be seriously damaged and out of action for many weeks, leading to significant city-wide disruption. Resilience implies a new concept of design for 'immediate occupancy' or 'functional recovery', terms intended to link safety with recovery time after a major earthquake (EERI 2019; Sattar et al. 2018). In both San Francisco and Los Angeles new legislation has adopted a resilience approach, and this has begun to be implemented (Mayor of Los Angeles 2016; Poland 2008).

Significant recent successes identified by our respondents also include massive expenditure on strengthening programmes to address vulnerable building types. This combined with the extensive replacement of the older building stock with new better-built buildings is seen as contributing to rapidly increasing resilience (Tobin 2019).

The California Geological Survey has also developed new urban maps of potential ground failure hazards (liquefaction, landslide and fault rupture), and cities now require developers to avoid these areas or otherwise mitigate the hazard (Tobin 2019).

The failures identified include the inability of smaller cities and suburban communities to adopt the same resilience concepts or retrofitting codes as large cities, and also the continuing vulnerability of existing private schools which are not covered by retrofit legislation.

A further problem is lack of earthquake insurance. Currently, only about 10% of California's housing is covered by the earthquake insurance offered by the California Earthquake Authority (CEA), and a similar proportion of commercial buildings, so a major event causing widespread damage would cause a financial crisis for many. However, for well-designed and constructed buildings the worst likely losses could be below the deductible level covered by the available insurance policies, and for many building owners not taking insurance is a calculated decision (Tobin 2019).

Although considerable progress has been made, many professionals think it is not enough and that a big earthquake occurring in the near future near a major city like San Francisco would still cause massive disruption and financial loss (Comerio 2020).

8.2.2 Japan

The whole of Japan is in a region of high seismicity, and its eastern coast, where most of its major cities lie, has a very high seismicity, as well as being at risk from tsunamis triggered by offshore earthquakes. As in California, progress towards earthquake protection in Japan has been stimulated by the experience of a series of major damaging earthquakes. The M7.5 Nobi earthquake in 1891, which caused over 7000 deaths, demonstrated that brick masonry was an unsuitable construction method for Japan, while the Mw8.1 1923 Great Kanto earthquake, with the terrible fire which followed in Tokyo, and a death toll of over 100000, showed both the strength and weakness of wood-frame construction. It is relatively resistant to ground shaking, but susceptible to fire. Two post-war earthquakes, the Mw6.6 Mikawa earthquake in 1945 and the Mw6.8 Fukui earthquake of 1948, each resulted in several thousand deaths, and in 1950, a completely new building code was adopted, which was implemented in the post-war reconstruction. In 1981, a major update of this

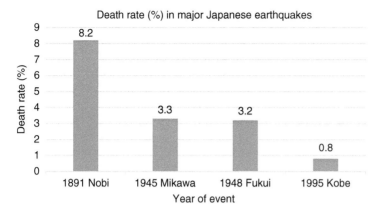

Figure 8.2 Reduction in death rate (proportion of population killed in the zone of highest intensity) in Japan in major earthquakes over 100 years. *Source:* After Y. Ohta, Unpublished data.

code was adopted, introducing ductility rules. These codes, and subsequent upgrades, have been effectively implemented. One way to measure the effectiveness of these codes is through fatality rates in major events: Figure 8.2 charts the very substantial reduction in the fatality rate in major earthquakes over a period of 100 years. As the chart shows, at the time of the 1891 Nobi earthquake, the death rate in the epicentral area was over 8%, while in the 1995 Kobe earthquake, it was less than 1%.

The Mw6.9 1995 Kobe or Great Hanshin earthquake nevertheless demonstrated the continuing vulnerability of the older wood-frame housing, as well as pre-1981 reinforced concrete construction. And in recent years, the Mw9.1 2011 Great Tohoku earthquake with its accompanying tsunami, which killed over 18 000, exposed many continuing weaknesses, particularly in tsunami protection.

Successes perceived by our respondents are the high level of adherence to building codes, the limited amount of building damage due to ground shaking in the Tohoku earthquake and the decrease in the number of fires over time, as a result of a nationwide implementation of gas shut-off devices. Japan leads the world in early warning after an earthquake has occurred, enabling rapid protection measures, for example gas shut-off, lift immobilisation and braking of high-speed trains. Although the Building Standards Act has not undergone a major revision to its earthquake resistance provisions since 2000, the current standard has been shown to be enough to provide protection against an event with ground-shaking level similar to that of 2011. Public policy has shifted towards soft measures for tsunami protection, such as better communication of early warning, evacuation drills and land-use planning (EEFIT 2013). A very high proportion of both schools and hospitals built prior to 1981 have been strengthened according to laws in place since 1995.

Failures included the lack of preparation for the 2011 tsunami, which was at a level significantly higher than was allowed for in the design of sea-defences and preparation planning, leading to the hugely costly and disruptive Fukushima Daiichi nuclear power plant failure. In the 2016 Kumamoto earthquake sequence, many buildings were destroyed because of the occurrence of two separate shocks of equally high intensity (Mw6.3 and Mw7.0) within two days, which is not accounted for in the codes. And although code adherence is still strong, building designers look to save costs by not adhering to recognised good practice in many aspects of building configuration and detailing (Okada 2019). Fatalities continue to occur through non-structural causes, for example, falling furniture or partition walls.

A further perceived failure is the continuing slow rate of strengthening of pre-1981 residential buildings. The government has set a target of 95% safe construction by 2020 with tax incentives for retrofitting work. However, by 2013, only 6 million of the 15 million pre-1981 dwellings had been strengthened (Kiyono 2019).

Due to the increasing demand for residential building plots, buildings are increasingly built on slopes or poor ground, and this has led to an increase (as in the 2018 Hokkaido earthquake) in the number of buildings damaged by landslides and liquefaction.

Social aspects of earthquake protection are also seen to need more attention. There is seen to be a lack of disaster prevention education, especially in areas not recently damaged by an earthquake. And present disaster risk prevention and recovery strategies are not thought to adequately cater for the needs of vulnerable social groups, the elderly, those with disabilities and foreign residents (Koyama 2019).

It is also worth noting that earthquake insurance in Japan is weak. The government only covers a small proportion (typically about 3%) of potential losses, and most householders have no additional cover.

8.2.3 New Zealand

As in Japan and California, it was the experience of a really devastating earthquake, the Mw7.4 Napier earthquake in 1931, which set New Zealand on the path to become one of the leading countries in earthquake protection worldwide. Much was achieved in the 80 years between that and the Canterbury earthquakes of 2010–2011, though with only a few damaging earthquakes to maintain the momentum. But the Canterbury events (Mw7.1 Darfield earthquake in December 2010, followed by Mw6.2 Christchurch earthquake in February 2011) and the Mw7.8 Kaikoura earthquake in 2016 have greatly increased the exposure of building and infrastructure owners and the public to the severity of the impacts of major earthquakes, particularly the effects on multiple properties simultaneously. Awareness amongst owners and the general public rose markedly as a result, and the experience of these events has led to some new initiatives and approaches to the design of buildings, the long-term benefits of which are still to be evaluated.

Successes as perceived at the time of the 2007 survey concerned the culture of research feeding into technical innovations and leading to improved codes and standards for building. The research at University of Canterbury on behaviour of reinforced concrete structures and leading to the concept of capacity design (Park and Paulay 1975) was seen as pivotal. New Zealand was also first to develop the lead-rubber base isolation system as a means to control earthquake damage, now widely adopted. And New Zealand was the first country to have adopted legislation (in 1968) enabling local authorities to identify 'earthquake-risk buildings' and require them to be strengthened (or demolished). This led to strengthening programmes in a number of places, notably Wellington, both for unreinforced masonry buildings and for non-ductile older reinforced concrete buildings. However, progress on this was slow, and many of these older buildings suffered extensive damage and were demolished following the Canterbury earthquakes.

Failures identified by the current survey respondents were the loss of most unreinforced masonry buildings in the 2011 Christchurch earthquake, and the associated loss of important architectural heritage; and also the demolition of a large number of damaged reinforced concrete buildings, which although they survived the earthquake with only moderate damage, could not be economically repaired (Figure 8.3). The large number of houses built on land which liquefied in the earthquakes and the extent of damage to such houses was also identified as a failure; as was the slowness and complexity of the insurance claims process (see Chapter 9). A further failure in the Mw7.8 2016 Kaikoura earthquake was the poor performance of precast concrete flooring systems in a number of multi-storey buildings in Wellington.

By contrast, the successes perceived by the survey respondents included the fact that almost all buildings in the 2011 Christchurch earthquake designed since the adoption of modern codes in the 1970s did meet life-safety requirements, even though the level of ground shaking was significantly higher than specified in the code. The same was true for the Kaikoura earthquake. Following the 2010/2011 earthquake sequence, there has been a dramatic increase in the awareness of the earthquake risk by the general public. A number of recommendations of the Royal Commission set up to investigate the Christchurch

Figure 8.3 Damaged reinforced concrete building in Christchurch after 2011 earthquake.
Source: Photo: author.

earthquake and its aftermath were implemented and have led to improvements in the building industry. Based on extensive multi-university research programmes, improvements have been made in national building standards. And new guidelines for assessing existing buildings for their earthquake risk have been published (NZSEE 2017) which will be used to target new strengthening programmes.

Another significant success for New Zealand is the very high proportion of buildings which are covered by insurance. Insurance with the EQC national insurance scheme is very high, and over 80% of all the losses caused by the 2011 Christchurch earthquake and 64% of those caused by the Kaikoura earthquake were insured (Table 2.1).

And notably, since 2011, there has been a very considerable increase in the application of resilience concepts in building design. In the case of New Zealand, this means the adoption of LDD, design of the structure of buildings for low damage in the design earthquake, and thus for rapid post-event occupancy (Cattanach 2018). This has been accompanied by the development of guidelines to frame performance criteria for LDD (Campbell 2018) to help reduce damage to non-structural elements of the building (cladding and services).

Thus New Zealand, already one of the world's most earthquake-aware countries before 2011, is now responding to the latest events in ways that should further increase the resilience of buildings in future earthquakes.

8.2.4 Chile

Located on the Pacific coast of South America, Chile is one of the most earthquake-prone countries in the world. The entire country is located in zones of high or very high seismicity. Since 1900 more than 80 earthquakes with a magnitude greater than 7 have been recorded. By a combination of earthquake experience, public awareness and effective government action, Chile has developed over the years a capacity to cope with earthquakes effectively, and is regarded as one of the world's most earthquake-resilient countries on account of its regularly updated and strictly enforced construction codes (Stein and Toda 2013). A measure of this was the relatively small death toll of 525 resulting from the Mw8.8 Maule earthquake in 2010, which also triggered a tsunami.

Successes of the last 10 years have been a direct response to the experience of that event. Those noted by our respondent include upgrading of the design codes to address topics not addressed before the 2010 earthquake, for example better design rules for reinforced concrete shear-wall structures, and new guidance for base-isolated structures. Also, there has been a much extended use of base isolation, following the excellent performance observed in base-isolated structures in the 2010 earthquake. All public hospitals built since 2010 have been required to incorporate seismic isolation. A new requirement to inspect the standard of construction of structural elements on site (not just through design checks) is also seen as a step forward in seismic protection.

The 2010 earthquake, which fortunately occurred outside of school hours, affected about one-third of all the country's public schools, and many school buildings were damaged. A programme of inspection, repair and retrofitting of 4000 affected schools was carried out over the next few years, creating a much better earthquake resistance than had existed at the time of the event (de la Llera et al. 2017).

A significant failure noted by our respondents is the absence of a requirement to consider earthquake hazard and risk in urban planning regulations in Chile, permitting urbanisation of land which is located on earthquake fault traces (Rivera et al. 2019).

As Rivera puts it 'Although Chile's approach to deal with the earthquake risk from a technical perspective via safe buildings has had overall successful results... the portion of the seismic risk which can be tackled by high-quality buildings is limited. Further attention needs to be put on urban and political discussions about planning, zoning, governance and institutional framework. Falling short in these dimensions of seismic risk may put Chile in a fragile condition for coping with its favourite natural hazard in the near future' (Rivera et al. 2019).

8.3   Limited Achievers

8.3.1   Greece

Over the last fifty years, the public's perception of earthquake risk in Greece has been alerted by frequent damaging events which have caused casualties. It was the devastating Mw6.8 Ionian Islands earthquake of August 1953, in which 476 people died, which led to the introduction of the first Greek earthquake code in 1959. Since 1978 several significant earthquakes have affected urban areas. The Mw6.4 Thessaloniki earthquake in 1978, the Mw6.7 Gulf of Corinth earthquake in 1981, the Mw6.0 Kalamata earthquake in 1986, the Mw6.5 Aeghion earthquake in 1995 and the Mw6.0 Athens earthquake in 1999 collectively caused 257 deaths, and considerable economic losses.

Greece's perceived successes are that as a result of these urban disasters, the public's perception of earthquake risk is probably the highest in Europe, and this has been accompanied by upgrades of the earthquake code (in 1986, 1995 and 2004), by improved building standards, and by government action. The Greek earthquake code is generally applied in new buildings, especially after the experiences of the 1978 and 1981 earthquakes. Rapid economic growth in the period 1960–2009 led to much new building. Almost all new buildings have been of reinforced concrete frame construction, and the use of *pilotis*, open ground floor on widely separated columns – creating a soft storey – a practice that was common from the 1970s until the late 1990s is now virtually discontinued (Figure 8.4).

The government funds several excellent universities supporting research and the training of structural engineers and in 1983, established the Earthquake Planning and Protection Organisation (EPPO) to plan and oversee a national policy for earthquake protection. EPPO has assembled and analysed data on earthquake damage and also set in place a programme of evaluating and upgrading the existing stock of public buildings. During 2004 to 2009 the Greek School Buildings Organisation assessed 33% of all school buildings and identified 9% needing immediate action. Strengthening works were carried out on a few highly vulnerable school buildings in areas of high earthquake hazard.

Perceived failures are that there continues to be a lack of formal and independent on-site inspection system during the construction of new buildings. Also Greece's sovereign debt crisis (since 2010) and ensuing austerity measures have been a major setback for earthquake risk reduction in Greece. There has been very little new building to replace the older

Figure 8.4 Reinforced concrete building on pilotis (Athens, Greece). This form of construction is now largely discontinued. *Source:* Photo: A. Pomonis. Reproduced with permission.

more vulnerable building stock, and building owners also find it difficult to maintain the older buildings, leading to a gradual increase in vulnerability.

A further consequence of the economic crisis is that funds have not been available for the process of assessing the vulnerability and carrying out necessary strengthening for some 80000 public buildings, and these programmes have virtually come to a standstill.

Greece therefore remains well-equipped in terms of both public awareness and available technical skills to make progress in earthquake protection, but lacks the financial resources to take the action required.

8.3.2 Italy

Substantial parts of Italy are regions of moderate to high seismicity, although, unlike Greece, its largest cities, Milan, Rome and Naples are in regions of lower seismicity. Development of codes and regulations for building to resist earthquakes started earlier in Italy than anywhere else in Europe, after shock of the Mw7.0 Messina (Sicily) earthquake of 1908, which killed over 80000 people. But the early seismic zonation applied only in the south of the country. The Mw6.5 earthquake of Friuli in 1976 and the Mw6.9 event in Irpinia in 1980, which killed 929 people and 4680 people, respectively, demonstrated the much greater proportion of the country at risk from earthquakes, and also exposed the high vulnerability of much of its building stock, including many buildings of cultural and historical importance.

Following these events, a national seismic zonation was developed, bringing more areas within the building code. But further events, Mw6.0 in Umbria and Marche in 1997 and

Mw5.8 in Molise in 2002 also caused widespread damage, though fewer casualties. The Molise event, however, caused the collapse of a school building which killed 26 schoolchildren and a teacher, triggering the start of a programme to strengthen schools in the highest-risk areas, as well as further modifications to the national building code. In the last decade, three further damaging events have taken place, the Mw6.3 2009 L'Aquila earthquake (309 deaths), the Mw6.0 2012 Emilia-Romagna earthquake (27 deaths) and the Mw6.2 Amatrice earthquake in 2016 (300 deaths).

Successes of the last 10 years identified by our respondents include the construction of emergency housing following the L'Aquila earthquake, a programme which had to be completed in six months before the onset of winter. Rather than the normal temporary housing (tents, containers, etc.), over 14 000 displaced L'Aquila residents were provided with accommodation built to a high standard, and including base-isolation anti-seismic construction, which was completely new to Italy at that time. These were to be used while their homes in the city were rebuilt, and later to be available for students or visitors (Figure 8.5). Set against that, however, must be the very long time needed for the complex reconstruction needed for the city centre buildings, which 10 years later, was still incomplete. Following the 2016 event, which involved four separate regions (Abruzzo, Lazio, March, Umbria), there was no such centralised provision of temporary housing, and reconstruction has again been very slow, with many bureaucratic obstacles, and inadequate financial support from the government.

Figure 8.5 CASE post-earthquake housing system used in L'Aquila earthquake. *Source:* Photo: Dolce and Di Bucci (2018). Reproduced with permission.

Italy has done much over the years to improve its emergency management procedures, which are today among the most sophisticated in the world. Worth particular mention is the care and protection provided for churches and other cultural heritage buildings, both through temporary strengthening of buildings and temporary care of artefacts. Another success has been the procedures which have been developed for the building-by-building assessment of damage, to assist in rehousing and assessing the costs of reconstruction. For example, within a year after the 2016 earthquake, over 210 000 building inspections had been carried out by expert assessors (Dolce and Di Bucci 2018).

Respondents also pointed to success in the developments in the national building code in recent years, bringing it more in line with the Eurocode (EC8). Changes include a point-by-point identification of seismic hazard over the whole country, replacing previous seismic zones and improved methods for inclusion of local soil response in the evaluation of seismic actions on structures. In Italy, the building code has the status of law, which means it must pass through Parliament. This is often a slow process but requires it to be used in all construction. Recent construction is thus expected to be safe against expected future earthquakes. But unfortunately, the current pace of construction is slow, and only 5% of homes in Italy were built in the last 20 years, so the legacy of unsafe buildings remains very large.

Some progress is being made in bringing the older building stock to an acceptable level of safety. In school building, a national programme is in place, but it is estimated by one respondent that only 10% of schools are currently safe. And since 2017, legal measures have been in place to offer tax incentives to owners to upgrade their buildings, with tax reduction rising with the level of seismic improvement achieved. It is too early to say how effective this measure will be.

Given the very large numbers of currently unsafe buildings, even where technically possible, the task of bringing all buildings to a safe level is immense and is estimated by one respondent at €93bn. Currently, the whole annual government programme of seismic protection is less than 1% of this figure. Thus, as our respondents state (Zuccaro and De Gregorio 2019): 'The path to reach the goal of securing the Italian territory with respect to the seismic risk is still very long'.

8.3.3 Romania

All of Romania's large earthquakes occur in the relatively small Vrancea region in the Carpathian Mountains, in an area which is not densely populated. But these are deep earthquakes, and as a result, they are felt over a wide area. They can be destructive to tall buildings at some distance from the epicentre in areas of deep alluvial soils, such as the capital Bucharest. Three major events occurred in the twentieth century. The first Mw7.8 event, in 1940, killed an estimated 1000 people. The most lethal event, Mw7.5 in 1977 killed nearly 1600, over 90% of them in Bucharest. This event also destroyed 33 000 dwellings, mostly in high-rise apartments, which were common by that date. A further smaller Mw7.0 shock occurred in 1990.

As a consequence of these earlier earthquakes, Romania has given considerable attention to earthquake-resistant construction, especially in Bucharest, and has developed an active and internationally well-integrated earthquake research community. There has not, however, been any earthquake resulting in building damage in the last 10 years.

Romania's perceived success has been in the development of a modern code of practice. First introduced in 1963, and regularly updated since then, the current code is now harmonised with the Eurocode (EC8). The complexity of the new design assessment rules has needed the upgrading of the skills of structural engineers, and government and the universities have been active in preparing local commentaries and setting up training programmes. Significant research in seismic protection for Romania has been conducted at the Technical University of Bucharest (UCTB), including a national seismic instrumentation network.

There have been upgrading programmes, supported by the World Bank, for public school and hospital buildings, and a programme is currently being formulated for the future upgrading of the national building stock to combine improvements in energy efficiency and earthquake risk reduction.

Romania's principal failure is perceived to be the significant residue of about 350 vulnerable buildings in Bucharest which have been identified as having the highest level of seismic risk, and on which strengthening is proceeding only slowly. Many of these are high-rise apartment buildings: some of them were damaged previously in the 1977 earthquake, but were only cosmetically repaired at the time (Figure 8.6).

Figure 8.6 Pre-1977 reinforced concrete apartment block in Bucharest, showing local repairs after the 1977 earthquake. *Source:* Photo E-S. Georgescu. Reproduced with permission.

Added to this is a perceived lowering of the standards of training of engineers and standards of design resulting from the shift to a market economy since the 1990s. According to Sandi (2019) what this amounts to is that 'a disaster prevention culture is not sufficiently developed in the country'.

8.3.4 Turkey

Much of Turkey's population and industry is located close to the highly active North Anatolian Fault or other adjacent fault systems, and during the twentieth century, Turkey experienced more earthquake disasters than any country other than China. The total life loss in the twentieth century was officially in excess of 80 000, and 12 events since 1900 have each caused more than 1000 deaths. In still recent memory, the events of 1999, the Mw7.6 Kocaeli earthquake at the eastern end of the Marmara Sea in August, followed by a Mw7.2 event at Düzce in November, between them caused widespread building collapse and over 18 000 deaths. These two events demonstrated how vulnerable the country's buildings were, particularly, and worryingly, those reinforced concrete apartment blocks most recently constructed.

The 1999 events, coupled with widely publicised studies which showed that an earthquake causing major damage to the huge metropolitan area of Istanbul is highly probable within 30 years (Parsons et al. 2000), have set in motion important changes in public awareness and government action on earthquake risk mitigation. Since 1999 Turkey has been one of the most active countries, and Istanbul one of the most active cities, in studying and trying to reduce its earthquake risk (Bogazici University et al. 2003).

One important successful step was the merging in 1999 of three separate agencies into the Disaster and Emergency Management Authority (AFAD), which now has a clear role of promoting disaster mitigation in addition to the government's traditional emergency management role. This has led to improvements in seismic hazard mapping and further updating of the earthquake code (to include seismic isolation). Establishing AFAD has led to the development of a National Earthquake Strategy and Action Plan (UDSEP), with specific goals to achieve by 2023 in the fields of research, hazard mapping, earthquake-safe settlements, education and public awareness and emergency management. A National Platform for Disaster Risk Reduction has also been created to draw all elements of Turkish society, NGOs, the private sector, universities and citizens to work alongside government towards risk reduction goals. However, progress in recent years has been hampered by the reduction in the status of AFAD within central government to an affiliated authority within the Ministry of the Interior, and by the Syrian refugee crisis which since 2011 has consumed most of AFAD's resources as the responsible government agency to provide a humanitarian response for some four million refugees.

A second area of success has been the establishment, in 2000, of a compulsory earthquake insurance pool (TCIP) for all new residential construction, providing insurance sufficient to cover basic reconstruction costs for a relatively small premium. Some nine million properties were, by 2019, covered by this insurance (see Chapter 9).

A further area of success is the enactment of a law (in 2001) which requires independent supervision of the design and construction of all new buildings, in all 81 provinces of the country, before the local authority issues its construction permit. The law was aimed to ensure that the required design standards were not ignored, either by designers or builders, as they had been in buildings destroyed in previous events. However, the actual implementation of the law and its enforcement is thought to be patchy.

In addition to these successes, there have been programmes, notably in Istanbul, for the assessment and strengthening of public buildings. The ISMEP programme in Istanbul, to assess and strengthen all seismically substandard schools, hospitals and administrative buildings, is scheduled to last until 2023 (ISMEP 2014) (Figure 8.7). Elsewhere, there are programmes for assessing schools and hospitals, and about 30 new city hospitals have been built to a high standard of earthquake protection including many with seismic isolation devices.

An important perceived failure is the lack of affordable retrofit schemes for existing residential buildings. Although skills and techniques for such work are available, there is little take-up, because the benefit achieved for the building owners is perceived to be less than the cost of the work.

Figure 8.7 Strengthened school building, the İhsan Hayriye Hürdoğan Elementary School, Istanbul. *Source:* ISMEP (2014).

As a result, Turkey currently has a huge number of buildings which would not survive the earthquake for which new buildings are required to be designed, perhaps 35% of the national building stock. The collapse of numerous apartment blocks in Izmir in the Mw7.0 Aegean Sea earthquake of 30 October 2020, with the loss of over 100 lives, is an indication that many people remain at serious risk. On the other hand, it is estimated that over the last two decades about half of Turkey's building stock has been replaced, and if this trend continues, and better design standards are implemented for the new buildings, the proportion of unsafe buildings will diminish quite rapidly over time.

8.3.5 Slovenia and Serbia

Earthquake-awareness in both these countries was stimulated by the Mw6.0 Skopje, North Macedonia earthquake of 1963, which killed 1100 people, followed in 1979 by the Mw6.9 Montenegro earthquake, which killed 136. At the time of these events both Slovenia and Serbia, with Macedonia and Montenegro, were republics of the former Yugoslavia.

In Slovenia, earthquake awareness was also stimulated by the Mw6.5 Friuli earthquake of 1976 in Italy but close to the border with Slovenia. In Serbia concern was triggered by the 2010 Mw5.3 Kralievo earthquake, which caused much more severe damage to structures than would have been expected in an earthquake of this magnitude.

A perceived success in both countries is that the earthquake code first introduced in 1963, was updated in 1981, and that Slovenia was the first country to adopt Eurocode 8 (EC8) (CEN 2004), which has been mandatory since 2008. Slovenia also has a relatively high knowledge of earthquake engineering among structural engineers. In contrast, Serbia still (in 2019) uses the 1981 Yugoslavia code.

A perceived failure in Slovenia is that not enough has been done to identify and strengthen older buildings (built before the code). This includes a number of 10–13 storey unreinforced masonry structures in the capital Ljubljana. Also that review of structural design by an independent authority is not required. Lack of a recent earthquake, as well as the demands of a market economy, is seen to make it difficult to maintain high standards of design.

In Serbia, the poor performance of buildings in the 2010 Kralievo earthquake was particularly noticed in masonry buildings (mostly unreinforced masonry) built before 1964 which had recently been upgraded, indicating failure to comply with the provisions of the regulations at the time. Some more modern reinforced concrete buildings were damaged, with similar conclusions. In neither country have there been programmes to identify and strengthen older highly vulnerable buildings, whether public buildings or residential buildings.

8.3.6 Portugal and Spain

Portugal and Spain are shown on the GEM Global Seismic Hazard map (GEM, 2018) as countries of mainly low seismicity, with some areas of moderate seismicity along the southern coast. Yet, in 1755, Portugal suffered the most catastrophic European earthquake of the last millennium. There is no scientific consensus on the origin or magnitude of that event, but the ground shaking was clearly sufficient to cause immense and widespread destruction, particularly around Lisbon, further aggravated by the effects of a massive tsunami. Lisbon's metropolitan area is also affected by earthquakes of relatively smaller magnitude originating in the Lower Tagus Valley, and in Spain, the southern region of Murcia was struck in 2011 by the Mw5.1 Lorca earthquake, which in spite of its moderate magnitude, caused significant building damage to the masonry building stock, and claimed nine lives. The Azores region of Portugal is also regularly damaged by moderate-sized earthquakes. The long experience of earthquakes in both countries has resulted in active scientific and earthquake engineering communities, well-funded through the EU and national and regional governments, and excellent programmes of training for earthquake engineering in the universities. However, it has been difficult to translate this into effective action for earthquake risk mitigation beyond the routine application of each country's earthquake code.

In Portugal, the most significant perceived success in recent years has been the preparation of the Portuguese version of the Eurocode (EC8) (CEN 2004), and the development of the National Application Document. This includes a new hazard map for continental Portugal and the Azores, and classification of soil types for the volcanic formations of the Azores.

The greatest failure was seen to be the lack of formal approval of these documents to enforce them as national code of practice, in spite of continuous active lobbying by the engineering community. Direct consequences are that the large amount of building upgrading that has taken place in recent years has been made without any agreed requirement for earthquake loading. In a few cases, the upgrading even increased the seismic vulnerability of the housing stock.

In Spain, the situation is similar. National Application Documents and hazard maps for application of EC8 have been developed, but government approval is still (2019) pending. The applicable code is out of date and fails to identify areas of significant hazard, such as that where the 2011 Lorca earthquake occurred.

In both countries, much of the building stock is old and highly vulnerable to even moderate earthquakes, but except to a limited extent in the Azores, there have been no sustained national programmes for strengthening of public buildings, schools and hospitals.

Activities to support public awareness occur regularly. In Portugal, the anniversaries of past earthquakes have been used to organise campaigns to communicate earthquake risk to the public in general and schoolchildren in particular.

8.3.7 Australia

Australia is a country with generally low seismicity, with only two small pockets of moderate seismicity, one around Adelaide, and the other in Western Australia. As a result, there were no provisions for earthquake loads in codes of practice for building until 1995, except in Adelaide, where they were introduced from 1981. The general introduction of earthquake codes was the result of the 1989 Mw5.4 Newcastle earthquake, in which 16 people died, and there was very widespread and costly building damage both in the city of Newcastle and much of New South Wales, particularly to the typical low-rise masonry buildings of the area (Figure 8.8). The event showed that even in areas of low hazard, damaging earthquakes can occur and will be more damaging if the buildings are built without any consideration for earthquakes.

Figure 8.8 Damage to low-rise masonry building in Newcastle, 1989 – still a concern in Australia. *Source:* Photo: EEFIT. Reproduced with permission.

Today, Australia has an active earthquake hazards research programmes at Geoscience Australia, and earthquake engineering research and teaching programmes at several universities.

Successes in earthquake protection identified by our survey respondent included the regular updating of the Australian earthquake design code and publication of a commentary for structural designers. Set against this, one failure is what our respondent describes as 'lack of understanding of basic seismic design principles within practising engineering design offices, which often results in the adoption of poor structural systems and detailing which lacks robustness or is inconsistent with ductility design assumptions' (Griffith 2019). With few significantly damaging earthquakes in Australia in the last 30 years, this situation is reinforced by public and client complacency towards the seismic hazard.

A standard also exists for retrofitting existing buildings for earthquake, but is not much used, although buildings undergoing significant renovation are required to meet current code standards.

Some strengthening programmes have taken place. In the 1990s, the South Australian state government made an assessment of its own building stock and identified 18 high-risk buildings which were seismically upgraded. And in the last decade, the Victoria and New South Wales state governments have been including seismic improvements in the programmes for maintenance of their most iconic heritage buildings.

8.3.8 China

Over several centuries, more lives have been lost from earthquakes in China than in any other country. The twentieth century, with 170 fatal events claiming over 600 000 lives, was no exception (Coburn and Spence 2002). For many decades, China has devoted much scientific effort to the earthquake problem. Its apparent success in earthquake prediction astonished the seismological world in 1975, when, following a swarm of small tremors, the small city of Yingkou in the Province of Haicheng was successfully evacuated shortly before a major Mw7.0 earthquake. The result of the evacuation was that there were many fewer casualties than might have occurred, given the level of building damage (Bolt 1999; Jones 2018). But the occurrence just over a year later of the Mw7.5 Tangshan earthquake, with the loss of about 250 000 lives, destroyed confidence in the earthquake prediction approach. The Tangshan event was not predicted and occurred in an area for which no specific earthquake-resistant design was at that time required.

Since then, efforts have been concentrated on the more conventional approach of earthquake fault and hazard mapping, developing and updating codes for new construction and retrofitting buildings in high-risk areas. However, disastrous earthquakes have continued to occur, most notably the 2008 Mw7.9 Wenchuan earthquake, which killed 87 000 people and the 2010 Mw6.9 Yushu earthquake, which killed 2700 people, demonstrating that China continues to be highly vulnerable to earthquakes. About one-third of China's territory is in areas of moderate to high seismicity.

Successes identified by our respondents have included updating the construction codes, along with a new zonation map, issued in 2015, including an increased protection requirement for schools and hospitals.

Principal perceived failures were the weaknesses in construction standards revealed by the very large number of buildings which collapsed in the Wenchuan and Yushu earthquakes, including many schools (Figures 2.12, 8.9).

In the Wenchuan earthquake, this was in part due to the fact that the ground shaking level in the earthquake in some places was substantially higher than design codes required. But corrupt evasion of the required standards by the builders has also been blamed, with much controversy (Wong 2008). Another related failure noted by our respondents is that most parts of the country, the earthquake design codes are not implemented in buildings built by private owners.

After each major earthquake, alongside reconstruction, there have been substantial programmes to retrofit public buildings, schools, hospitals and university dormitories in

Figure 8.9 Partially collapsed five-storey restrained masonry building in the 2008 Wenchuan earthquake. *Source:* Photo: EERI. Reproduced with permission.

the areas affected, and there have been projects in several regions to improve construction of rural buildings.

The city of Hong Kong is a special case. Although in a zone of low seismicity, a 2004 report by consultants Arup indicated that there is a significant risk of damage to unreinforced masonry buildings of less than seven storeys. A code of practice based on the Chinese code combined with elements of the Eurocode is currently being prepared and should be implemented in 2020. More than 60% of the Hong Kong building stock consists of buildings over 15 storeys, and the 2004 report importantly indicated that there is low risk to such buildings, mainly because of their robust design to resist typhoon winds which are common in Hong Kong.

8.3.9 Mexico

Spurred by the Mw8.0 Mexico City earthquake of 1985 which caused enormous building damage and life loss in the capital city, Mexico's earthquake engineering research and training of engineers is at a high level. Published codes and standards are of international standard and regularly updated.

But survey respondents report that there is no adequate mechanism for enforcement of standards and no site supervision. There is also little by way of land-use planning regulations, and corruption is widespread. As a result many supposedly engineered buildings are substandard and vulnerable to earthquakes. And most of the population of the country (both urban and rural) live in non-engineered buildings. Some of these are of unreinforced masonry or adobe and are highly vulnerable to earthquakes. However, there is a strong tradition, dating back many years, of using confined masonry, without the involvement of engineers, and buildings built this way performed relatively well in the Mw7.1 Puebla earthquake in 2017 (Reinosa 2019). A programme supported by the World Bank for upgrading schools is at an early stage.

8.4 Continuing and Growing Risks

8.4.1 Colombia and Ecuador

Colombia is located in a highly active earthquake zone. Major cities, Bogota, Medellin and Cali as well as other medium-sized ones, lie in zones of high or very high seismicity. Over the last 500 years, more than 50 earthquakes of magnitude greater than 7 have hit Colombia. The most recent disastrous earthquake was the Mw6.1 Quindio event in 1999, which killed 1900 people, but several events since then have also caused casualties. Colombia's long earthquake experience has been an important stimulus to national efforts to reduce earthquake risk. This has not only involved developing and updating a national building code in line with the international state-of-the-art but also efforts to develop and promote appropriate guidance documents for use in non-engineered buildings.

Perceived success of the last 10 years have been the further updating (2010) of the earthquake resistant construction code, including updating of the seismic hazard, seismic microzonation of several cities and enhanced standards for essential buildings. In 2012, the public policy on disaster risk management in the country was updated, using a new approach that focussed on risk reduction, not just post-disaster management. As a result of this policy, several cities have now carried out building-by-building seismic risk modelling to guide strengthening programmes and insurance. There is evidence that the seismic risk in some cities has been reduced because of improved compliance with standards in the formal sector.

The main weaknesses identified by our respondent are in the areas of enforcement of standards in new construction and programmes for strengthening existing buildings. Although there has been new legislation to improve technical supervision in the construction phase, which has had an effect on the formal construction sector, most construction in the cities is in the informal sector and not affected by the legislation. And even though the high vulnerability of the marginal settlements which continue to increase in the country's towns and cities is well known, there has been a lack of recent media attention to the earthquake problem.

Colombia has nevertheless had a successful programme of evaluating and strengthening of school and university buildings in particular cities (Cardona 2019), notably Bogota (Figure 8.10), and there are plans to extend this to other cities and also to hospitals, but progress on a national level has been slow or non-existent.

Neighbouring Ecuador has a comparable level of seismic risk, affecting particularly the capital city Quito. A destructive Mw7.8 earthquake occurred in the coastal Muisne-Pedernales area in 2016, resulting in over 600 deaths. The extent of the disaster was identified as being due to a failure to implement earthquake-resistant design codes and regulations and lack of building control. The post-disaster management, organised by the army, was

Figure 8.10 Strengthened University building in Bogota, Colombia. *Source:* Photo: O. Cardona. Reproduced with permission.

seen as a success (EEFIT 2016). However, the earthquake has had little lasting effect on construction standards, and there have as yet been no programmes for strengthening of high-risk public buildings.

In both of these countries, in spite of some success, the words of (Cardona 2019) continue to be true:

> it is necessary to be more radical in requesting effectiveness and commitment. If we reinforce one school but we need to intervene in 10000 the achievement is nothing. The problem grows faster than the velocity of the solutions: this is the main failure of our social commitment as academics and professionals.

8.4.2 Indonesia

The islands of Indonesia lie within one of the most active earthquakes zones in the world. More than 150 earthquakes larger than M7 have occurred in Indonesia in the period 1901–2017, and more than 25 earthquakes have each killed more than 10 people since 2000. Some of these earthquakes have their epicentre offshore and trigger tsunamis, which can greatly increase the number of fatalities. On 26 December 2004, the massive Mw9.1 Indian Ocean earthquake and tsunami killed an officially reported 167000 people in Indonesia alone, mostly through the impact of the tsunami on coastal settlements. But most of the deaths in recent earthquakes, such as that in the Mw7.5 2018 earthquake in Sulawesi, have occurred through the collapse of buildings not built to withstand earthquakes. Like other earthquake-prone countries, Indonesia has developed and updated a modern code of practice for buildings, which is based on the equivalent US codes, and this is used for major constructions in the big cities and for public buildings. However, the vast majority of building construction, both in the cities and in the rural areas, is non-engineered construction, built without the involvement of professionals and without the use of any earthquake code (Boen 2016).

Successes noted by our respondent include the development of the code of practice for construction, and its enforcement in the major cities (Jakarta, Medan, Surabaya, Semarang). In these cities, most of the high-rise buildings and infrastructure are designed by certified engineers and to a good standard.

A further success is the research into non-engineered construction, leading to the development of manuals for non-engineered construction appropriate for the forms of construction used in different parts of the country. Research has also been carried out on methods for the strengthening of existing non-engineered buildings using techniques which are simple, affordable and replicable, and these have been applied in a number of projects (Figure 8.11) (Boen 2016).

The massive internationally financed reconstruction programme which took place in Aceh province after the 2004 tsunami, might in some respects be described as a success. But later observations (Arup 2006; Boen 2008) suggested that much of the construction was of very poor quality and that many of the techniques were based on imported materials and techniques, which are not replicable by the local people. After the reconstruction, it was observed that people have reverted to previous highly vulnerable building methods.

Other failures identified include the non-application of the code of practice for earthquake-resistant design, even for engineered buildings, because city authorities (other than those mentioned above) do not make it mandatory; failure to apply widely the available manuals for non-engineered construction; a lack of public awareness on the extent of the earthquake risk and of the techniques available for safer housing construction; and a lack of any campaigns to promote such awareness.

Another failure noted is a decline in the capability of building artisans and technicians compared with 20 years ago, and a lack of interest by trained professionals (engineers and architects) to learn about non-engineered construction methods. Although (in Indonesia as elsewhere) many NGOs have become involved in projects to build local artisan capacity, the combined impact of such programmes is still small.

The consequences are that earthquakes continue to result in avoidable building collapses and loss of life, as shown in the recent earthquakes in 2016 in Aceh region (Mw6.5, over 100 deaths), in 2018 in Lombok (Mw6.9, over 500 deaths) and also in 2018 in Sulawesi (Mw7.5, over 4000 reported deaths, some due to tsunami).

Figure 8.11 Techniques for strengthening non-engineered buildings in Indonesia. *Source:* From Boen (2016). Reproduced with permission.

According to (Boen 2016):

> Despite the many human casualties and severe impact on the regional economy and development, it seems that relatively little effort/action is being done to prepare, prevent or mitigate the effects of future earthquakes. All of the damages to date are repetitions of past occurrences and are a demonstration that in Indonesia not much improvement has been made with regard to non-engineered construction.

8.4.3 Nepal

Nepal is a small country in the Himalayan region of high seismicity and has a history of devastating earthquakes. In 1934, the huge M8.0 Bihar–Nepal earthquake caused very strong ground shaking in the Kathmandu Valley, where the capital is located, destroying 20% of the valley's building stock. The urbanisation of the last 50 years has further concentrated population in the valley, with declining standards of construction, and this has led to a major concentration of risk. In a 2001 study, Kathmandu was ranked most at risk out of 20 cities assessed globally (Bothara et al. 2018). However, the Mw6.9 East Nepal earthquake in 1988, and more recently, the Mw7.8 2015 Gorkha earthquake caused serious ground shaking over a wide area, and this has led to major efforts to increase public awareness and reduce the risks. Over the last 10 years, a series of programmes have been conducted, notably the Kathmandu Valley Earthquake Risk Management Project and the School Earthquake Safety Project discussed in Chapter 7 (Figure 7.2).

Raised earthquake awareness and enhanced risk perception throughout the country is regarded by our respondent as the most significant success achieved in the Nepal in the last decade. In the view of this respondent, 'Nepal has conspicuously departed from the fatalistic view of disasters as being an Act of God'. This has been accompanied by the success of the school building retrofitting project, improved implementation of building codes, improved national emergency response capacity and a community-based disaster risk management programme. This has enabled the government to declare the building code mandatory for the entire country including rural areas.

As a result, positive changes have taken place in the urbanising municipalities in the implementation of simplified codes for non-engineered buildings. In 30 urban municipalities outside of Kathmandu valley, code compliance increased dramatically, from 14 to 74% within a span of five years, when technical assistance was provided for the municipalities.

Failures include the fact that, in the country as a whole, an estimated 65% of rural buildings and 35% or urban buildings are still built without applying codes. There is still seen to be a lack of effort towards building technical capacity for earthquake-resistant construction and for checking compliance. Although seismic assessments of hospitals have been in progress for some years, little hospital retrofitting has been done. There is also seen to be a serious gap in research into appropriate methods to improve local technologies for construction and for protection of historical and religious monuments. And disaster risk reduction is not yet adequately integrated into the economic development process.

Nepal's efforts at disaster risk reduction were severely tested in the Mw7.8 Gorkha earthquake of 2015, which caused massive damage to rural housing, and caused nearly 9000 deaths (Chapter 1). There was an effective disaster management process, and much of the recent building stock survived undamaged, including all of the 300 school buildings which had been retrofitted in the School Earthquake Safety Programme, none of which was severely damaged. But the fact that over 880 000 houses were affected indicates that Nepal still has much to do. A recent review (Bothara et al. 2018) concludes that

> The earthquake risk in Nepal is rapidly increasing due to the combined effects of seismic hazard, earthquake vulnerability of the building stock and infrastructure, limited investment, limited political will to implement building-related legislation and codes, fatalism, and a weak economy.

8.4.4 India

The parts of India with the highest seismicity are the northern states in or bordering the Himalayas, Jammu and Kashmir, Himachal Pradesh, Uttar Pradesh, Bihar and Assam and neighbouring states, but there is also an area of high seismicity in the state of Gujarat, and moderate seismicity in much of northern and western India. India has been shaken by two particularly devastating earthquakes in recent times, the Mw6.2 Latur earthquake of 1993, which caused over 7000 deaths, and the Mw7.7 Bhuj earthquake of 2001, which killed 13 800 people. In addition, in the tsunami triggered by the Mw9.1 Indian Ocean earthquake, 10 000 people died in India. These events revealed the vulnerability of much of India's rural housing. But the collapse of numerous modern structures in the Bhuj earthquake, in particular a number in the city of Ahmedabad, 230 km away from the epicentre (Figure 8.12) has indicated the very high level of earthquake risk that exists in much of urban India too. As one example, the city of Delhi, currently home to 20 million and growing rapidly, is on the edge of the Himalayan zone of high seismicity. These recent events have been a catalyst to important actions on the part of the Indian government to improve earthquake risk management.

One primary set of successes noted by our respondents has been legal and administrative steps to create the framework for a new national system for disaster management, which shifts the emphasis from response to preparedness and mitigation. A National Disaster Management Agency has been established, with parallel agencies at State and District levels, and a national Disaster Response Force to deal with disasters or threatening disasters. New model building byelaws have been drawn up to improve building standards and to

Figure 8.12 Partially collapsed building in Ahmedabad, 230 km from epicentre of M7.7 Bhuj earthquake. *Source:* EEFIT (2005). Reproduced with permission.

ensure a system of building control at the level of states and cities, and building standards have been updated.

In addition, national funding for research on seismology and earthquake risks has been increased, and the course curricula for training of students of civil engineering and architecture in earthquake safety have been improved. There have been some programmes for the training of building artisans (Chapter 7). All of these activities are likely to be beneficial to earthquake safety in the long term, and there has been some observed improvement in the extent of implementation of earthquake codes of practice, primarily in the formal sector in urban areas (Sinha 2019).

Failures identified are that engineering is not a legally regulated profession in India, as a result of which there are inadequate means to enforce the professional conduct of engineers. A high proportion of professionals also do not have the knowledge and skills to implement existing safety standards. Even in urban areas, there is still a widespread lack of compliance with standards, development control regulations and building byelaws and many substandard buildings continue to be built. In rural areas, few buildings are built with any consideration of earthquake safety, and as cities expand, these become part of the urban building stock (Jain 2016). There are a few mason training programmes (Chapter 7), but these are organised by individual NGOs and not part of an institutionalised culture of safe building practices.

India has a vast and rapidly growing population. According to the 2011 census, there were over 330 million housing units in India. Two-thirds of these were in rural areas and predominantly constructed from weak masonry materials, while many of those in the rapidly growing cities are constructed without adequate safety in design or building control. In the words of Sudhir Jain (2016):

> In the years ahead, India will have a construction boom in housing and infrastructure of staggering magnitude... Given current construction practices and the lack of an adequate regulatory framework to manage such construction, the seismic risk in the country is growing rapidly.

8.4.5 The Caribbean

Most of the Caribbean islands have moderate-to-high seismicity, but until 2010, the previous 50 years had been relatively free from major earthquake disasters. Going further back, a Mw7.5 event in the Dominican Republic in 1946 caused 2550 deaths and a M6.5 event near Kingston in Jamaica in 1907 claimed about 1000 lives, so the risks were known to specialists. But the earthquake hazard was not well-understood either by the general public, or, in much of the Caribbean, by the engineering profession. Although standards existed on some islands, they were not well implemented.

So the occurrence of the Mw7.0 event affecting the poverty-stricken capital city of Port-au-Prince in Haiti, with the loss of an estimated 220 000 lives, the world's most lethal earthquake in the twenty-first century, was a shock. But this event has done much to galvanise public opinion and government action not just in Haiti, but in the other Caribbean territories too.

Successes noted by our respondent included new studies of the seismic hazard in the eastern Caribbean region as well as for Jamaica and Cuba, which will help put earthquake-resistant design on a sounder basis. An increasing percentage (though still less than 50%)

of engineers in the region are consciously doing aseismic design. Earthquake codes are present in most Caribbean countries but commonly implemented mainly in Dominican Republic, Puerto Rico, Jamaica, Trinidad and Tobago and the French Antilles. And base isolation techniques have begun to be implemented in the French island of Martinique, though not yet elsewhere.

Since 2007, there has also been a major project to investigate and strengthen hospitals throughout the region, supported by the Pan-American Health Organisation. A large-scale screening programme has been carried out, and a project to retrofit 50 healthcare facilities in seven Caribbean countries is currently underway. A few Caribbean countries also have programmes for the assessment and upgrading of schools and other public buildings, notably in the French Antilles (Gueguen 2019).

To set against these successes, even where design of buildings takes account of the earthquake loads, detailing for ductility is usually absent. Architects continue to lack interest in appropriate design concepts for earthquake resistance, and electrical and mechanical plant is regularly designed without consideration for earthquake loads. Worryingly, our respondent considers that there is almost no progress towards effective enforcement of the earthquake-resistant standards.

Thus, while buildings built to modern codes are likely to avoid serious damage in a major earthquake, much of the building stock in the Caribbean countries is unsafe, and according to our respondent (Gibbs 2019) 'the majority of buildings would be severely damaged if an earthquake of the severity prescribed in the current codes were to occur'. This includes most of the housing stock, where unsafe unreinforced masonry building methods are common. The reinforcing practices for masonry vary considerably from island to island. In Jamaica, the walls are always reinforced. In Trinidad and Tobago, they are almost never reinforced. For most of the other islands, the practice falls somewhere between these extremes (Gibbs 2019).

8.4.6 Other Countries with Growing Risks

Within the same earthquake belt as India and Nepal, and both with a high seismicity over much of the country, Pakistan and Myanmar have become more aware of their earthquake problem in recent years and have taken some action to mitigate the disaster risk. In 2005, Pakistan was shaken by the Mw7.6 Kashmir earthquake, which resulted in 87 000 deaths. There was a huge internationally supported reconstruction programme in the years following the event, as a result of which some improved housing and schools were built in the area and public awareness of the earthquake risk was enhanced nationally. The Pakistan Building Code was also changed in 2007 to upgrade earthquake provisions. But means to enforce and implement this code are still lacking. Most of the country's building stock, in both urban and rural areas, remains vulnerable.

Myanmar also has a history of damaging earthquakes, and its government has recently begun to look for ways to increase earthquake preparedness and response, through a number of UN-supported studies. These include a proposed new building code, a national earthquake preparedness and response plan, an earthquake risk and vulnerability assessment for Yangon City, and a primary school curriculum of earthquake dos and don'ts. But earthquakes are widely seen as an Act of God, and priority in limited government expenditure is given to other basic services and to cyclone and flood protection. To date, there has been little action to build safer structures.

In Africa, responses were received from two countries, Mozambique and Ghana, each of which has some regions of moderate seismic risk, and some experience of damaging earthquakes. In Mozambique, there was a large Mw7 earthquake in 2006 in which damage was largely confined to rural areas. This has not been followed by any risk reduction measures because of the infrequency of earthquakes compared with other natural disasters such as floods, cyclones and droughts, and because of the competition for investment in other basic services, healthcare, schools, electricity and water.

In Ghana, parts of which are regularly shaken by moderate earthquakes, a *National Building Guide for lightly loaded structures in disaster-prone areas* was published in 2011 (Tetty 2011), with provision for earthquake loadings, and a committee was appointed in 2019 to develop a comprehensive framework for earthquake preparedness and response. But little has been done to enforce the National Building Guide. Western Accra, the most earthquake-prone part of the country, has been the site of much building in recent years, but with no earthquake protection measures.

8.5 Country Comparison of Unsafe Structures

One of the survey questions asked respondents to estimate the number of unsafe structures (pre-code buildings or those not built to a code) in urban and rural buildings, using their own personal judgement. A number of respondents were unable to give a response to this question, and several gave a response which did not distinguish between urban and rural buildings. In a few cases, some interpretation by the authors of the survey responses was made to create an estimate. The resulting estimates are shown in Figure 8.13, in which the countries have been collected into three groups as in the discussion above.

The results are closely in line with the classification. In all the countries in the 'high achievers' group, the proportion of unsafe buildings is low, between 5% and a maximum

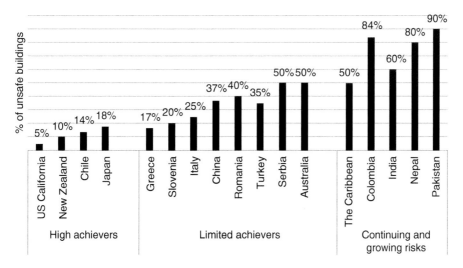

Figure 8.13 Country comparison of estimated proportion of unsafe structures, organised by country groups.

of 18%. In the 'limited' achievers group, the proportion of unsafe buildings varies from 20% to a maximum of 50% (in both cases of 50% unsafe buildings, these are countries of only moderate or low seismicity, Serbia and Australia). In the countries with continuing risks, the estimates given by our respondents are all greater than 50%, with some estimates as high as 85% (Colombia) and 90% (Pakistan). Although a definition of what constitutes an unsafe building is difficult to give precisely, and interpretations vary, the results shown in Figure 8.13 do serve to give some estimation of the current level of earthquake protection achieved, and to help to define the task remaining to be done.

8.6 Comparison of the Country Groups

8.6.1 High Achievers

The four countries or regions we categorise as 'high achievers' (Japan, New Zealand, the US State of California and Chile) are all countries with a high per capita income (by World Bank definitions), and a highly urbanised population (>80%). Importantly, all have regular experience of damaging earthquakes, and thus they all have a high level of public awareness of the earthquake risk, and a correspondingly low public tolerance for extended interruptions caused by earthquake damage (Comerio 2020). All of them also have modern building codes, maintained by well-funded research programmes, and effective code enforcement. Professional education is good, and this is partly responsible for a high level of code compliance.

But recent experience and risk studies have shown that there is still potential for very substantial losses from expected earthquakes, even where the codes are effective. This is because the codes which are largely designed to create life-safety in the event of likely earthquakes do not prevent damage which could cause severe disruption and economic losses. This has led to the growing adoption of a resilience concept of earthquake protection, where earthquake risk is viewed at community rather than building-by-building scale, and codes are designed to allow for immediate post-earthquake occupancy. A further aspect of the resilience approach is to identify older buildings whose failure could cause life-loss and disruption and require these to be strengthened, with legally enforceable timescales for their retrofitting. Such an approach, only so far adopted by a few cities, is increasingly seen as the model for the future.

A further success of these countries in the growing adoption of advanced technologies such as base-isolation in large buildings to limit the impact of future earthquakes.

Remaining failures of the high-achiever countries are that adoption of the resilience concept tends to occur city by city, and thus smaller cities tend to be left behind; insurance cover, though strong in New Zealand, is surprisingly limited in Japan and California. With local exceptions, though schools and hospitals are now mostly dealt with in these countries, the rate of strengthening of the older at-risk homes and workplace buildings is very slow, leaving much of the population still at risk. And, mainly in Japan, much still needs to be done to provide protection of the coastal population from tsunamis.

8.6.2 Limited Achievers

These countries include both high-income countries (most of Europe) and a few middle-income countries (Romania, China, Turkey), with urban population percentages varying from 50 to 90%. In all of these countries, modern earthquake design codes and regulations exist, engineers are trained to implement them, and these codes are for the most part thought to be implemented in recently constructed buildings. However, in all of these countries, there are a large number of older buildings, built without use of earthquake design codes, or built using now outdated codes, including many public buildings and buildings of historic importance, which are highly vulnerable to expected earthquakes.

In a few of these countries, where there is recent experience of damaging earthquakes (China, Greece, Turkey), public awareness of the earthquake risk is high, but in most of them, lack of recent damaging earthquake experience means that public concern, and pressure for earthquake protection action, is low. In several countries, there have been limited campaigns for strengthening public buildings such as schools and hospitals, but in most of them little or nothing has been done to tackle this task, let alone the much greater task of strengthening the general residential and commercial building stock.

In several countries, such as Greece and Romania, although a structure exists at government level for taking such action, at least for public buildings, the country's economic situation has hampered progress.

Other failings noted are the lengthy delays in several countries in reconstruction following recent earthquakes, and residual weaknesses in the implementation of the code, due to lack of independent inspection.

8.6.3 Continuing and Growing Risks

Even though many of the countries in this group have experienced devastating earthquakes in recent years, all of them are middle or low-income countries with substantial rural populations, and construction of most dwellings continues to be highly vulnerable to earthquakes.

Even though construction in many of these countries is nominally governed by earthquake codes, these codes only apply to the formal construction of urban areas, and even there code enforcement is frequently inadequate or non-existent, and most construction does not meet adequate safety levels. In the rural areas (where a substantial part of the population lives), most construction uses various forms of masonry, normally without any consideration of earthquake safety.

In many of these countries, there is a group of dedicated earthquake engineering professionals, and important work has been done in developing codes and guidelines, not only for formal sector construction but also for non-engineered construction, and in demonstrating better building techniques through artisan training programmes. The popularity of these programmes proves their value, but to date, these efforts have only reached a small fraction of the populations at risk.

There have also, in several of these countries, been important campaigns for the improvement of safety in schools and hospitals (e.g. in Colombia and Nepal), and these have been demonstrated to be effective. But as a result of low family incomes, the homes of the majority of the population are still being built using conventional methods using available local materials, and even when aware of the earthquake risk, any resources homeowners can find to build better tend to be used to create more living space or other improvements.

Urbanisation is occurring rapidly in many of these countries, but moving to urban areas generally means finding space in highly vulnerable informal settlements, or building in a similar way to the construction of rural areas. Buildings in modern materials, if not built to earthquake codes, are not necessarily any less vulnerable.

Thus for the large and often rapidly growing populations of these countries, the earthquake risk today is increasing rather than diminishing and is creating the context for large earthquake disasters in the future.

Acknowledgements

This chapter was based substantially on the survey responses of country and regional experts. The authors are very grateful to all our respondents and apologise for any remaining inadequacies in the representation of their replies. Thanks to Mary Comerio, Tom Tobin, Tom O'Rourke and Charles Scawthorn (California), Junji Kiyono and Shigeyuki Okada, Charles Scawthorn, Maki Koyama and Anawat Suppasri (Japan), Andrew Charleson and David Hopkins (New Zealand), Felipe Rivera and Juan Carlos de la Llera (Chile), Antonios Pomonis (Greece), Giulio Zuccaro, Daniela de Gregorio and Agostino Goretti (Italy), Emil-Sever Georgescu (Romania), Polat Gülkan (Turkey), Peter Fajfar (Slovenia), Svetlana Brzev and Borko Bulajić (Serbia), Carlos Oliveira (Portugal), Amadeo Benavent (Spain), Philippe Gueguen (France and French Antilles), Michael Griffiths (Australia), Zenping Wen and Xun Guo (China), Jack Pappin (Hong Kong), Omar-Dario Cardona (Colombia), Eduardo Reinoso and Mario Ordaz (Mexico), Manuel Querembas (Ecuador), Teddy Boen (Indonesia), Amod Dixit (Nepal), Ravi Sinha and Sudhir Jain (India), Tony Gibbs (Caribbean), Harosh Lodi (Pakistan) and Titus Kuuyour (Ghana, Mozambique, Myanmar).

References

Arup (2006). *Aceh & Nias Post Tsunami Reconstruction – Review of Aceh Housing Program.* London: Ove Arup and Partners.

Boen, T. (2008). Reconstruction of houses in Aceh, three years after the December 26 2004 tsunami. *Presented at the International Conference on Earthquake Engineering and Disaster Mitigation 2008.*

Boen, T. (2016). *Learning from Earthquake Damage: Non-Engineered Construction in Indonesia.* Indonesia: Gadjah Mada University Press.

Bogazici University, Istanbul Technical University, Middle East Technical University, and Yildiz Technical University (2003). *Earthquake Master Plan for Istanbul.* Construction Directorate, Metropolitan Municipality of Istanbul.

Bolt, B. (1999). *Earthquakes,* 4e. W.H. Freeman and Company.

Bothara, J., Ingham, J., and Dizhur, D. (2018). Earthquake risk reduction efforts in Nepal. In: *Integrating Disaster Science and Management* (eds. P. Samui, D. Kim and C. Ghosh). Elsevier.

Campbell, P. (2018). Proposed low damage design guidance – a NZ approach. *Presented at the 17th US–Japan–New Zealand Workshop on the Improvement of Structural Engineering and Resilience* 2018, Queenstown, New Zealand.

Cardona, O.-D. (2019). Survey response.

Cattanach, A. (2018). 12 projects over 12 years: reflections from implementing low damage designs. *Presented at the 17th US-Japan-New Zealand Workshop on the Improvement of Structural Engineering and Resilience* 2018.

CEN (2004). EN 1998: design of structures for earthquake resistance (Eurocode 8): European Committee on Standards, European Commission, Brussels

Coburn, A. and Spence, R. (2002). *Earthquake Protection*. Wiley.

Comerio, M. (2020). Survey response.

Dolce, M. and Di Bucci, D. (2018). The 2016–2017 Central Apennines seismic sequence: analogies and differences with recent Italian earthquakes. In: *Recent Advances in Earthquake Engineering in Europe* (ed. K. Pitilakis). Springer.

EEFIT (2005). *The Bhuj, India Earthquake of 26th January 2001*. UK: EEFIT, Institution of Structural Engineers.

EEFIT (2013). Recovery Two Years After the 2011 Tohoku Earthquake and Tsunami: A Return Mission Report by EEFIT. Institution of Structural Engineers.

EEFIT (2016). *The Mw7.8 Ecuador Earthquake of 16 April 2016: A Field Report by EEFIT*. UK: EEFIT, Institution of Structural Engineers.

EERI (2019). Functional recovery: a conceptual framework. www.eeri.org/wp-content/uploads/EERI-Functional-Recovery-Conceptual-Framework-White-Paper-201912.pdf (accessed October 2020).

Gibbs, T. (2019). Survey response.

Griffith, M. (2019). Survey response.

Gueguen, P. (2019). Survey response.

ISMEP (2014). *Retrofitting and Reconstruction Works*. Istanbul Project Coordination Unit.

Jain, S.K. (2016). Earthquake safety in India: achievements, challenges and opportunities. *Bull. Earthq. Eng.* 14: 1337–1436.

Jones, L. (2018). *The Big Ones: How Natural Disasters Have Shaped Us, and What We Can Do About Them*. Doubleday.

Jones, L. and Benthien, M. (2011). Preparing for a "Big One": the Great Southern California shakeout. *Earthq. Spectra* 27: 575–595.

Kiyono, J. (2019). Survey response.

Koyama, M. (2019). Survey response.

de la Llera, J., Rivera, F., Mitrani-Reiser, J. et al. (2017). Data collection after the 2010 Maule earthquake in Chile. *Bull. Earthq. Eng.* 15: 555–588.

Mayor of Los Angeles (2016). *Resilience by Design*. City of Los Angeles.

NZSEE (2017). *The Seismic Assessment of Existing Buildings: Technical Guidelines for Engineering Assessments. Overview and Summary*. New Zealand Society for Earthquake Engineering.

Okada, S. (2019). Survey response.

Park, R. and Paulay, T. (1975). *Reinforced Concrete Structures*. Wiley.

Parsons, T., Toda, S., Stein, R.S. et al. (2000). Heightened odds of large earthquakes near Istanbul: an interaction-based probability calculation. *Science* 288: 661–665.

Poland, C. (2008). *The Resilient City: Defining what San Francisco Needs from its Seismic Mitigation Policies.* San Francisco Planning and Urban Research Association (SPUR).

Reinosa, E. (2019). Survey response.

Rivera, F., Rossetto, T., Bardet, J., 2019. Understanding earthquake reslience in Chile: the Pros and Cons of safe buildings. *Presented at the SECED Conference* 2019, Greenwich, London.

Sandi, H. (2019). Survey response.

Sattar, S., McAllister, K., Johnson, K. et al. (2018). *Research Needs to Support Immediate Occupancy Building Performance Objective Following Natural Hazard Events*, NIST Special Publication 1224. National Institute for Standards and Technology, US Department of Commerce https://nvlpubs.nist.gov/nistpubs/SpecialPublications/NIST.SP.1224.pdf (accessed July 2020).

Sinha, R. (2019). Survey response.

Spence, R. (2007). Saving lives in earthquakes: successes and failures in seismic protection since 1960. *Bull. Earthq. Eng.* 5: 139–251.

Stein, R.S. and Toda, S. (2013). Megacity megaquakes – two near misses. *Science* 341: 850–852.

Tetty, J. (2011). *National Building Guide for Lightly Loaded Structures in Disaster Prone Areas.* Ghana: National Disaster Management Organization.

Tobin, T. (2019). Survey response.

Wong, E. (2008). Chinese stifle grieving parents' protest of Shoddy School construction. *New York Times.*

Zuccaro, G. and De Gregorio, D. (2019). Survey response.

9

The Way Forward: What Part Can Different Actors Play?

9.1 International Agencies and Global Initiatives

There are numerous international agencies that have disaster and, particularly earthquake risk reduction in their remit. Loosely these organisations can be divided into bi- and multilateral donor groups such as the International Development Funding in the United Kingdom, International Financial Institutions such as the World Bank and UN Agencies. There are also grant-based groups and projects, such as the Global Facility for Disaster Risk Reduction (GFDRR), and those programmes responding to global challenges, such as those set out by the Sendai Disaster Risk Reduction Framework (2015–2030).

Different international agencies have different geographical remits and reach. For example, UN Office for Disaster Risk Reduction (UNDRR) has a global remit, whereas the UN Development Program (UNDP), a national one. The types and levels, as well as location of activities carried out by the different agencies, would depend largely on the funding, regional and national developmental challenges and priorities.

In this section, an overview of some of the international organisations involved in earthquake risk reduction is highlighted. The summary will not include all agencies and programs involved in earthquake protection but will provide a flavour of the types of projects and information that is available and highlight some potential areas that warrant attention and funding in the future.

The UNDRR 'brings governments, partners and communities together to reduce disaster risk and losses to ensure a safer, sustainable future'. One of their recent initiatives is the Sendai Disaster Risk Reduction Framework which sets out a programme for the substantial reduction of disaster risk and losses in lives. The Sendai Framework lays out clear responsibilities, targets and priorities for reducing global disaster risks. In all, seven global targets are set out in this framework (2015–2030) as shown in Figure 9.1, which was agreed in Sendai in 2015, the location of one of the major natural disasters in recent times.

The Sendai Framework is designed to complement and work with other global 2030 Agenda agreements, including the Paris Agreement on Climate Change, the Addis Ababa Action Agenda on Financing for Development and the New Urban Agenda. Central to the thinking behind the 2030 Agenda agreements is that in order to achieve each of the programme targets, the approaches need to be connected: working cross-sectors, between and

Why Do Buildings Collapse in Earthquakes?: Building for Safety in Seismic Areas,
First Edition. Robin Spence and Emily So.
© 2021 John Wiley & Sons Ltd. Published 2021 by John Wiley & Sons Ltd.

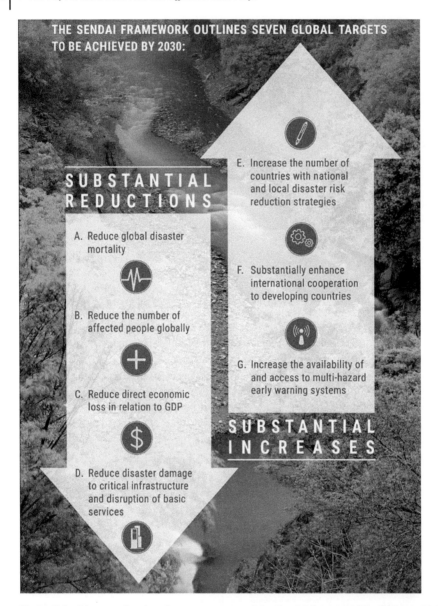

THE SENDAI FRAMEWORK OUTLINES SEVEN GLOBAL TARGETS TO BE ACHIEVED BY 2030:

SUBSTANTIAL REDUCTIONS

A. Reduce global disaster mortality

B. Reduce the number of affected people globally

C. Reduce direct economic loss in relation to GDP

D. Reduce disaster damage to critical infrastructure and disruption of basic services

E. Increase the number of countries with national and local disaster risk reduction strategies

F. Substantially enhance international cooperation to developing countries

G. Increase the availability of and access to multi-hazard early warning systems

SUBSTANTIAL INCREASES

Figure 9.1 Diagram showing the seven targets of the Sendai Framework for Disaster Risk Reduction set out by UNDRR in 2015. *Source:* The Sendai Framework Outlines Seven Global Targets To Be Achieved By 2030, What is the Sendai Framework for Disaster Risk Reduction?. United Nations Office for Disaster Risk Reduction.

within institutions, and ensuring harmony from policy through to activity. In the words of the Head of UNDRR, Mami Mizutori, 'Nothing undermines development like a disaster'. Haiti is a prime example where a nation regresses and becomes stuck in a vicious cycle of 'disaster > response > dependency > repeat'. As such, there is a clear alignment between the ambitions under UNDRR and UNDP in progressing the Sendai Framework and meeting the UN's internationally agreed Sustainable Development Goals at the same time.

In addition to working with participating nations on setting out programmes under the Sendai agreement, there are other fora organised through UNDRR and by its member states to share knowledge. The Global Platform for Disaster Risk Reduction is a biennial multi-stakeholder forum established by the UN General Assembly to review progress, share knowledge and discuss the latest developments and trends in reducing disaster risk. In its commitment to 'leaving no one behind', the UNDRR has also set up several online platforms to share experiences and ensure equitable access to the latest disaster-related information. Complementing the global platforms there are also regional platforms that foster south-south knowledge exchange. At the community level, the PreventionWeb platform is one of the world's most referred global knowledge platforms on disaster risk and resilience.

Another related knowledge sharing partnership is the International Recovery Platform (IRP) working to strengthen knowledge and share experiences and lessons on building back better in recovery, rehabilitation and reconstruction. It is a joint initiative of UN organisations, international financial institutions, national and local governments, and non-governmental organizations engaged in disaster recovery and seeking to transform disasters into opportunities for sustainable development.

Furthermore, international funders also have special branches dedicated to overseeing projects related to risk mitigation and financing, these include the Global Facility for Disaster Risk Reduction (GFDRR) and Disaster Risk Financing and Insurance (DRFI) Program at the World Bank, the InsuResilience Investment Fund (IIF) backed by the German government and others, the Centre for Disaster Protection jointly funded by the previous two funders with UK AID and the Insurance Development Forum, and the Integrated Disaster Risk Management Fund through the Asian Development Bank, to name but a few.

The Asian Disaster Risk Center (ADRC) focuses on capacity building amongst its member states, much like the work of the GFDRR which is a trust fund facility at the World Bank. They work together with disaster-prone countries to acquire knowledge to better understand their risks and build local capabilities for managing those risks. Areas of work at the GFDRR include risk identification, supporting open data for resilience strategies, financial protection schemes and recovery resilience. As summarised in their Knowledge and Learning Catalog, GFDRR's work to date can be categorised into the following eight areas:

1) Using Science and Innovation in Disaster Risk Management
2) Promoting Resilient Infrastructure
3) Scaling Up Engagements for the Resilience of Cities
4) Strengthening Hydromet and Early Warning Systems
5) Deepening Financial Protection
6) Building Social Resilience
7) Addressing Climate Risk and Promoting Resilience to Climate Change
8) Enabling Resilient Recovery

Notable earthquake-related programmes include the Global Program for Safer Schools (GPSS), the GRADE framework for assessing damage post-earthquakes and the Understanding Risk (UR) forum, bringing together the global community of experts and practitioners with interest in the field of disaster risk identification, specifically risk assessment and risk communication.

In other global regions, the European Commission hosts a Disaster Risk Management Knowledge Centre and its European Civil Protection and Humanitarian Operations (DG ECHO) has the responsibility of improving resilience by reducing disaster risk in humanitarian action. Its Peer Review Program in disaster risk management and civil protection systems provides a country or a region with the unique opportunity to reflect on its readiness to cope with natural and man-made disasters and identify ways of strengthening its broader prevention and preparedness system. The Disaster Risk Management Network overseen by the Inter-American Development Bank provides a platform for innovative technical information to be shared between its member states.

One other initiative worthy of a mention is the Global Earthquake Model (GEM) which since its inception in 2009 by seismologists and engineers has brought the latest advancements in earthquake science and hazard and risk evaluation to many countries around the world. As shown in Figure 9.2, at its core is OpenQuake (OQ) comprising the engine, platform and tools that cater to a variety of users, from modellers and researchers to emergency planners.

Two of GEM's most successful products to date are the global hazard and risk maps, offering standardised comparative assessments of hazard and risk (costs) for all seismic countries. Underlying these maps are standardised datasets and methods of calculation for all countries to review and implement. Its mission is to become one of the world's most complete sources of risk resources and a globally accepted standard for seismic risk assessment, and to ensure that its products are applied in earthquake risk management worldwide.

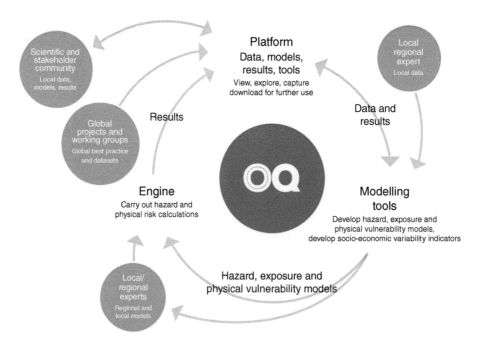

Figure 9.2 Visual showing the contributions by GEM's OpenQuake program. *Source:* Openquake, GLOBAL EARTHQUAKE MODEL. © GEM. CC BY-SA 4.0.

These programmes are very much needed as we strive as a global community for an equitable DRR future. With the advent of the Sendai Framework and other city resilience programmes, there has certainly been a shift in attention and resources to earthquake risk reduction. However, there are definite issues with sustainability of initiatives within local contexts that such international partnerships cannot achieve. The legacy of their work would rely on the engagement of local communities and action by governments.

9.2 Governments

A national government's role can be split into that conducted by the central government, national agencies and involvement at a local level with local agencies and city mayors. Obviously, every country has its own governance organisational structure and it is hard to generalise. Therefore, in this section, some examples of national and local initiatives that have made an impact on earthquake preparedness, especially in reducing vulnerability of the built environment, are highlighted.

The OECD Toolkit for risk governance lists 19 good practices for earthquakes risk reduction (OECD 2020). The OECD works together with governments, policy-makers and citizens, to establish evidence-based international standards and solutions to a range of social, economic and environmental challenges, providing a unique forum for sharing best practices.

Some of the good practices listed under earthquakes include regulatory instruments to restrict land use at a municipal level (in France); natural hazards information portals to share latest event data and research and analysis to the public and government agencies (Switzerland, France, Austria, and the United Kingdom); national risk estimation software (HAZUS_MH, USA, R-FONDEN, Mexico); federal funds for the prevention of natural disasters (Australia, Mexico); hospital and schools safety programmes (Mexico); national and regional education programmes and earthquake drills (Japan, USA); presidential declarations for mobilisation of resources for emergency response and recovery (Chile, Italy, and the USA).

As shown some of the good practices involve allocation of funding of earthquake risk mitigation and emergency response, others are regulatory in nature. These practices may of course not be feasible in all earthquake-prone countries; however, what is interesting to note from this list are the range and scope of government-led programmes, and their origins. In some cases, the regulations and policies have been implemented as a result of a catastrophic national event and special funding has been commissioned after the disaster. In other cases, the high seismicity of the region offers a constant reminder of the threat the population is under, and earthquake preparedness is therefore integral within national agendas.

Governments are needed to pass laws and assistance programmes to prepare for earthquakes. The challenge in government is that there are many risks to manage. These include more frequent natural risks such as floods and extreme weather, credit, political, financial, terrorism, biological, trade, climate change, fire, cyber-security risks, and more recently health risks. Governments will always be limited in the resources they have to reduce risks and furthermore, government terms are very short term relative to return periods of earthquakes.

To keep the reduction of risk to earthquakes on a government's agenda, foresight is needed to help build earthquake resistant communities. Individual advocates within governments and major national events are often the main reasons for change. In many countries, science advisory groups are in place to provide information and advice to governments, like SAGE in the United Kingdom, but they have no power and the final decisions and policy changes are ultimately made by the politicians, albeit 'following the science'.

The underrepresentation of scientists within governments was noted by Nick Smith, the Minister for the Environment and Building and Construction in New Zealand until 2017. He was the only scientist in the New Zealand Cabinet, and he cited this as a potential barrier to making political decisions in the field of earthquake preparedness (Smith 2017).

From a policy perspective, it is relatively easy to pass laws and regulations on newly constructed buildings, and codes are constantly upgraded with lessons learnt from actual events. It is a far more difficult to impose upgrades on existing buildings. Building control and regulations are crucial. In developed countries, it is the old buildings built before the development of seismic design standards, particularly those of unreinforced masonry that are responsible for deaths from earthquakes around the world.

Most seismically active countries have no legal requirements for older buildings to be upgraded. Taking New Zealand as an example, after the Canterbury Earthquakes and discovering that all 68 of its councils have different definitions and methodologies for upgrading, the government ordered a nationally consistent approach to retrofit and upgrades, and passed a law that allowed varying timeframes for buildings to be assessed and upgraded, relative to its earthquake risk. For example, in high-risk areas like Wellington, upgrades must be done within 15 years and in low-risk areas like Auckland, 35 years. These were in direct response to the Mw7.0 Darfield earthquake in 2010, the Mw6.1 Christchurch earthquake in 2011 and the Mw7.8 Kaikoura earthquake in 2016.

Legal requirements to strengthen old buildings should be one of the most important items on a government's agenda of improving seismic resistance in cities around the world.

9.2.1 Maintaining Professional Standards

A commitment and mechanism for maintaining professional standards in countries at risk is also needed.

In 2016, the New Zealand government made changes to the code of ethics that requires by law for engineers to pass on information deemed in the interest to public safety to relevant public authorities. This was a direct consequence of the collapse of the CTV building in 2011, in which 115 people were killed. When the building which was constructed in 1986 was up for sale in 1993, it was assessed by consulting engineers as deficient in its seismic design. The client wisely opted not to buy the building on this advice, but this information was not officially recorded nor passed on to the Christchurch City Council. The building was not sold, and the subsequent tenant CTV was not made aware of this safety concern either.

In countries such as Turkey and Japan, laws in place have enabled criminal charges to be brought against architects, engineers and builders who have failed to adhere to building codes.

A building contractor was tried for allegedly killing 195 people in the Mw7.6 Kocaeli earthquake in 1999, after 500 of the 3000 buildings he constructed in Yalova in Turkey collapsed. He was sentenced to 18 years and 9 months in prison. However, after only serving 7.5 years of this sentence, he was released in 2011 and has apparently reopened the construction and land office company (Erdin 2018). In 2005, a Japanese architect pleaded guilty to fabricating data leaving nearly 100 apartment buildings and hotels vulnerable to even modest earthquakes. He had falsified documents to cut costs for contractors for fear of losing their business.

Such prosecutions are rare and in mass fatality events such as Mw7.6 Kashmir earthquake in 2005, Mw7.9 Wenchuan earthquake in 2008 and Mw7.0 Haiti earthquake in 2010, there has been no accountability. It can be said the international and national construction industries have failed to protect the affected population.

To take action against badly designed buildings, investments by the government in education and training of professional engineers are needed. In addition, in countries such as Pakistan and India, there are also issues with retaining expertise. The issue of brain drain is prevalent across many developing countries. Many professionals choose to leave their home countries after acquiring their professional qualifications to work abroad, to the Middle East or Western Europe; or decide not to return after obtaining their education abroad. This seepage of skills and knowledge is detrimental to nations not only in economic terms but in public health and safety. Funding in training and retaining professional engineers must be prioritised.

9.2.2 Funding Scientific Research

Funding scientific research and innovation is another important remit for national governments. In the United Kingdom, the Global Challenges Research Fund (GCRF) was established in 2015 to unite international partners in support of new research into disaster risk, reduction and resilience. The £1.5bn programme funds advanced research that addresses the challenges faced by developing countries. Other research programmes specifically targeting disaster and earthquake risks include European Commission's Horizon 2020, the US National Earthquake Hazards Reduction Program (NEHRP), and Japan's Science and Technology Agency (JST), etc.

In countries where there is simply a lack of capabilities, for example many small Pacific Island nations and countries in Central Asia, requests for help from international agencies and learning from neighbours are important. Following the Mw6.1 Bhutan earthquake in 2009, which claimed 12 lives and caused an estimated loss of about US$54mn, a post-disaster needs assessment led by the government with support from the World Bank, the United Nations and the EU, highlighted that the Kingdom needed to understand its exposure to seismic risks better. Since then Bhutan has taken a multi-departmental approach to align the activities of each technical agency to improve the country's seismic resilience. This is vital as available resources and technical capacity in the country are limited. Since 2009, with the financial support of the Japan Policy for Human Resources Development (PHRD) and the World Bank, the first six earthquake monitoring stations in the country were installed. The *Improving Resilience to Seismic Risk* project funded by PHRD from 2013 to 2017 directed the creation of manuals and guidelines on seismic building construction. Working with its neighbour Nepal, seismologists from both countries, France and Switzerland have also established research into better understanding the seismicity of the Kingdom.

The Chinese Earthquake Authority (CEA) is a prime example of how changes have been implemented in traditional practices in the monitoring and assessment of earthquake hazard and risk. In the past, earthquake prediction was pursued by many seismologists in China, especially after the Mw7.0 Haicheng earthquake in 1975. With the advances of science and sharing of knowledge across the world, the CEA have been implementing changes to its focus and reprioritising their resources. Although the investigations into earthquake prediction is still a research arm within the agency, increasing monitoring networks, assessing building exposure and loss estimation models have become as important. Collaborations within and outside of China with academic and national institutions have resulted in a marked change in the organisational structure of CEA and through co-funded research schemes like the UK–China Research and Innovation Partnership Fund, earthquake professionals have been able to share experiences and learn from the CEA.

One recent success has resulted in a public awareness campaign jointly led by the CEA and academics in the UK and USA. The project ran from 2016 to 2019 and as called the Pan-participatory Assessment and Governance of Earthquake Risks in the Ordos Area (PAGER-O). We were involved in the project along with UK Overseas Development Institute (ODI), University of Oxford, Hong Kong Polytechnic University and Geohazards International as well as international risk consultants. The project brought together natural and social scientists, communications specialists and creatives with policy-makers, practitioners and local communities. At the beginning of the project, a series of charrettes were carried out in Weinan and Xian in China. The charrettes focused on understanding the main earthquake risk problems in Linwei and Huazhou districts. These were then written up as proposals and agendas for action together on site with the local teams. Throughout the project, transdisciplinary approaches were used to identify and fill knowledge gaps, and co-produce evidence-based approaches to reduce risk and increase resilience to earthquake hazards. The highlight and outcome of the project which was chosen by the local disaster management experts as the most attractive method of risk dissemination, was a story.

The *Homecoming* shown in Figure 9.3 is a story about how a strong earthquake affects a family with 'left-behind' children. This illustrated book won the first-class award (of book series) in 'The 2020 International Contest for Science Communication on Earthquake Preparedness and Disaster Reduction' organised by the Seismological Society of China. The story, which is in Chinese and English, is based on a scientific earthquake scenario of a repeat of the historic 1558 earthquake in the Weinan region in China. This is the aftershock of the deadliest recorded earthquake which was accompanied by numerous landslides in Shaanxi, China, in 1556. When the world's population was perhaps about 500 million, it is estimated that 830 000 people perished in that one event in 1556.

Through this three-year project, we produced a storytelling and graphic novel format used to help the public and decision-makers visualise an earthquake for the first time in China, with physical scientists, engineers, social scientists and artists collaborating on the scenario. The *Homecoming* story integrated local knowledge with technical findings and physical impacts with emotional consequences as well. Raising public awareness through different forms of media and in culturally and contextually relatable forms is central to earthquake communication and education.

Figure 9.3   Cover page of the Homecoming cartoon, produced as part of a joint UK and China research council funded project called PAGER-O. *Source:* Author's own project.

The Shannxi and Weinan earthquake authorities have been advocating the use of this material and a most promising outcome is that Weinan Education Administration believes that the storytelling and illustration-led narrative could be included as a school-based textbook or extra-curricular reading material for their safety education course. This would cover their whole prefecture-wide primary and high schools, which would include nine counties or county-level administration units, reaching out to over 500 000 students.

More recently at a global level, the Global Program for Safer Schools (GPSS) with support from GFDRR have developed Virtual Reality simulation and video sharing experiences in earthquake safety in California with school children across the world.

9.2.3 Education

The role of governments promoting disaster education in the national curriculum cannot be overestimated. When children are provided with an opportunity to understand hazards and identify local vulnerabilities and capacities, they can play a central role in disaster risk reduction and resilience. In the PAGER-O project, we focused on creating narratives as our key scenario products, because innovative disaster education material is very limited not only in the project study area, but also in the whole nation. China's educational ethos includes a long-established idea in safety education called 'big hand in small hand' (i.e. parents learn from their children) and studies have shown the importance of involving children's participation in DRR (Pfefferbaum et al. 2018).

As noted in the list of best practices for earthquake risk reduction by the OECD, Japan has one of the most comprehensive national curricula promoting earthquake safety. Several pedagogical tools are recommended, including storybooks, cartoons, games and the Internet, as shown in Figure 9.4.

Other methods of education are reminders embedded in everyday items such as the *furosihki*, traditional Japanese wrapping cloth traditionally used to transport clothes, gifts or lunches. The design shown in Figure 9.5 in English and Japanese has been developed by the International Research Institute of Disaster Science (IRIDeS) after the Mw9.1 Great Tohoku earthquake of 2011.

In recent years, there have been a shift towards more holistic, place-based participatory approaches to school-based disaster risk reduction and resilience education (DRRRE).

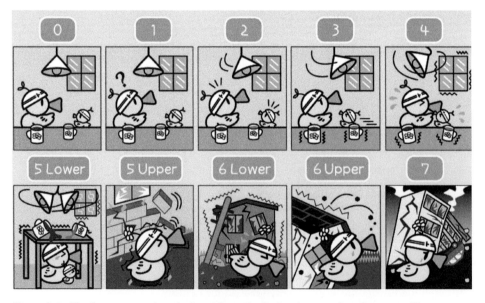

Figure 9.4 The Japanese earthquake intensity scale depicted as a series of cartoons. The illustrations show the Japanese intensity scale, from level 0 (no shaking that can be felt) to level 7, which brings widespread devastation. *Source:* Introducing how to deal with earthquakes in an easy-to-understand manner for foreigners with illustrations and English, the Japan Weather Association "Tokyo Disaster Prevention" website. © Japan Weather Association.

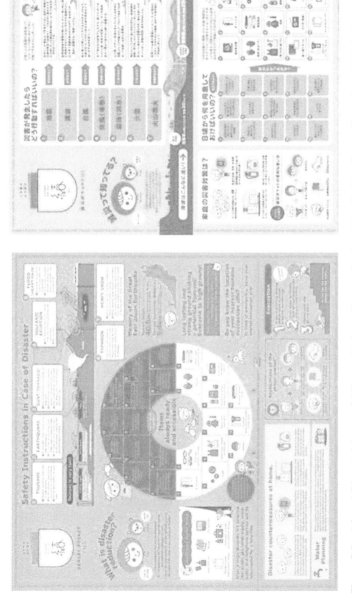

Figure 9.5 A furoshiki containing useful tips and information on disaster risk reduction. *Source:* Hazard and Risk Evaluation Research Division Tsunami Engineering, International research institue of disaster.

The core idea is to create relatable learning environments where students are actively engaged with problem-solving and not just reading from textbook examples. The three principles are the following:

- Understanding Local Hazards
- Identifying Vulnerabilities and Capacities
- Engaging in an Authentic Project

This hands-on approach has been piloted in countries such as Australia in their wildfire education program for 10–12-year-old children with positive results, including increased knowledge and awareness, increased household planning and preparedness, and increased child participation in a range of DRR and resilience activities at home and in the wider community. A participatory approach is also central to GPSS's work with schoolchildren across the globe, with a potential reach of millions.

9.2.4 Working with the Private Sector

Collapses of old unreinforced buildings are the main causes of injuries and deaths in earthquakes. Some of these will be public buildings but most will have single or multiple owners. Working with the building owners and businesses is essential to significantly reduce the exposure of these assets to earthquakes and save lives.

Again, returning to New Zealand, their government passed an Order in Council in 2017 requiring all earthquake-prone unreinforced masonry building owners with street facing facades and parapets in high occupation areas to tie back these features in the seismic areas of Wellington, Blenheim and Lower Hutt. The out of plane failure of these features killed 39 pedestrians after the Mw6.1 Christchurch earthquake in 2011. A time limit of a year was set for these changes to be implemented and a fund of NZD$3mn was earmarked to assist with this cost which, combined with councils, equated to a dollar for dollar subsidy. Subsidies such as these and other financial incentives like tax reductions from the national governments and local councils to encourage buildings and business owners to take on responsibilities to reduce earthquake damage and casualties are instrumental to action.

At a local city level, the *Resilience by Design* program launched in Los Angeles in 2014 to address earthquake vulnerabilities is perhaps one to follow, reviewing individual buildings and responsibilities as well as upgrades to public utility networks. During thorough consultations with earth scientists, the earthquake engineering and public policy professionals and hundreds of community organisations in the city, it became apparent that there was a disconnect of what is understood about earthquake risk, and most importantly, what can be done about these risks. The city of Los Angeles is encouraged to consider earthquakes as a societal issue. For example, since the creation of building codes, the onus has been on the owners of these buildings to implement the code. If an owner chooses to build a building that is a total financial loss, that is his/her right, if no one is killed. The reality of a major earthquake is that the failure of a single building would impact the whole community through economic disruption, population decreases and cascading failures of engineered and social systems. It is therefore in a city's interest to consider the realities of earthquake losses and insist on shared economic decisions. Since 2014, Northern California have followed suit and launched the Resilient by Design Bay Area Challenge, inviting designers to think not only about earthquakes but also to suggest solutions for other natural threats as well, such as flooding.

9.3 Businesses and Organisations

An earthquake can have a major impact on a private company or manufacturing organisation some of whose facilities are affected by the ground shaking or other hazards. Business managers have a dual responsibility, both for the safety of their employees and other visitors to their facilities in the event of an earthquake and for the continuation of their operations during and following an event. They can take numerous measures to protect the users of their buildings and to safeguard the operations and minimise disruption and are generally legally required to provide protection for their employees. The same is true for other organisations with responsibility for buildings such as schools, universities and medical facilities. Responsibility of employers can also extend to a concern for the earthquake safety of their employees' homes, which could prevent them from working if damaged.

9.3.1 Risk Assessment

The first action needed is to be fully aware of the hazards, and the same assessment as that described for homeowners in Section 9.4 below is applicable also for businesses. The hazard assessment must include each separate building and facility which the business is responsible for (whether as owner or tenant). Not only ground shaking hazard but also subsoil conditions, landslide and tsunami potential and other possible hazards must be considered, based on locally available information on the risk.

The vulnerability of each separate building and facility must then be considered. The assessment should consider not only the potential damage from both the largest earthquake shaking during the building's life (or in 500 years), as shown on seismic hazard maps, but also the annual expected loss from earthquakes of all sizes. For businesses or organisations with multiple facilities in different seismic regions, a preliminary assessment based on the building construction types may be used to identify locations which should have a high priority for protective action. The high-risk facilities should then be subject to a more detailed vulnerability assessment based on on-site evaluation by a structural engineer.

For locations at which buildings may collapse or be seriously damaged in the largest earthquake scenario, business organisations should also undertake casualty modelling studies to assess likelihood of casualties, death and serious injury, to building occupants and passers-by. Studies have repeatedly shown that casualty rates are much higher in some types of building than others, and Emily So's recent work (So 2016) has made it possible to estimate lethality and injury rates for different classes of building, based on the expected loss of volume in an earthquake scenario (Chapter 6). As shown in several recent earthquakes (for example the Mw7.8 Kaikoura earthquake in 2016), non-residential buildings can be responsible for a high proportion of the casualties when an earthquake occurs during working hours. In the neighbourhood of unreinforced masonry buildings on busy urban streets, the public can be at risk from falling parapets, chimneys and tiles. Taking action to protect against avoidable casualties should have a high priority in risk mitigation for businesses.

Another major concern for business owners is to avoid business interruption caused by a major earthquake. This is likely to be more costly than the exposure to direct damage to buildings (Yanev and Thompson 2008). Items of equipment within the premises need to be protected by firm attachment to their bases, and storage racks and other building contents protected. Business interruption may also occur if the local transportation networks are interrupted or homes damaged, and staff cannot get to work, or if vital telecommunications networks are disrupted. Supply chains may also be disrupted so that materials and supplies needed for manufacturing operations cannot reach a factory or outputs reach their destinations. Because of its unpredictability, assessing the potential for business interruption is more difficult than assessing direct damage or potential casualties.

9.3.2 Mitigation Plan

Having assessed the risks, a mitigation plan needs to be drawn up. There are in general three options: to accept the risks, to take mitigation action or to transfer the risk through insurance. Mitigation action needed may be no more than tying down equipment and securing the contents. Or it may involve retrofitting or even replacing existing buildings or moving to new premises. Design for resilience, which means ensuring the level of design is sufficient to ensure continuing operation immediately after a major earthquake, is increasingly being implemented by businesses and other organisations in the high-risk cities of the USA, Japan and New Zealand (Cattanach 2018), (Chapter 8). In a resilient city, the responsibility of business owners is seen to include minimising disruption to the city and not only to their own operations.

These options will need to be carefully evaluated based on detailed engineering assessment of what is required for each building. Insurance policies are likely to be available to cover building and contents damage costs, at an annual premium which will generally be assessed based on the specifics of the business. And insurance policies for business interruption may be available as an add-on to building and contents damage cover. Insurance policies come with a deductible which the business owner pays in the event of a loss. But the costs and benefits of the three options should be carefully assessed before deciding which to pursue. It is important to bear in mind that no retrofit strategy can eliminate the risk of some damage in a major earthquake.

9.3.2.1 Safe Workplace
Businesses also need to be active in training all employees on the actions to take in the event of an earthquake. They should encourage employees to take part on annual drills such the California Shakeout (Chapter 7), in which nearly 70 million people globally took part in 2019. The SCEC document Seven Steps to a Disaster-Resilient Workplace (SCEC 2016) identifies six categories of business assets: people, data, buildings, inventory, equipment and operations, and suggests measures for protecting each from any of 12 different hazards (natural or man-made).

Its recommendations for securing your space include securing equipment, moving heavy items onto lower shelves, not having open shelves near exits, and keeping space under desks clear for 'drop, cover and hold' actions during earthquake ground shaking. It also recommends that all business premises should prepare disaster plans including an evacuation process, storing emergency supplies, training some staff as emergency responders and plans for post-event recovery, as well as making sure that all staff are regularly updated on those plans. FEMA (2020) has provided similar recommendations.

9.4 Homeowners and Individual Citizens

9.4.1 Safe Home

Homeowners bear the principal direct responsibility for ensuring an earthquake-safe society. Whether as owner-occupiers or landlords, it is their understanding of the problem and their decision-making that determines whether residential buildings survive an earthquake or are damaged, whether the occupants are liable to injury and how well the occupants are able to recover. Residential buildings everywhere constitute the majority of the building stock, and their homes are where citizens are most likely to be at the unpredictable moment of occurrence of an earthquake.

Peter Yanev and Andrew Thompson, based on over 30 years of experience, have set out the basic requirements for achieving earthquake safety in their book *Peace of Mind in Earthquake Country: How to Save your Home, Business and Life* (Yanev and Thompson 2008). They propose a three-phase plan, with eleven steps, which homeowners should follow. Although primarily addressed to California citizens, the approach, with some modification, is applicable in earthquake country anywhere in the world.

The three phases are

- Phase 1: Assess your earthquake risk
- Phase 2: Decide how to manage your earthquake risk
- Phase 3: Implement your decisions

The principal component of the earthquake risk is the expected level of ground shaking from the strongest earthquake which can be expected. Earthquake hazard maps are now available for every country in the world. The Global Earthquake Model (GEM) (http://www.globalquakemodel.org) has prepared hazard maps, both globally and by region, and national geological survey departments also provide similar hazards maps, usually showing ground-shaking level which, based on current knowledge, would be exceeded once in 50 years and once in 500 years. These hazard maps are also the basis for the codes of practice for the design of new buildings.

In addition to the expected level of ground shaking, it is important to know about the subsoil conditions at the location of the building. Soft soil, as explained in Chapter 4, can significantly increase the level of ground shaking locally, and it is important also to know if the building is in an area subject to liquefaction (often a risk in low-lying coastal areas) or subsidence. In addition it is important to consider the potential landslide risk: landslides have been responsible for many building collapses and fatalities in recent earthquakes as detailed in Chapter 2. Many municipalities now publish land-use maps which identify areas of potential subsoil problems and landslide risk.

The presence of active faults near to a building should also be considered, since surface fault ruptures can be devastating for buildings close to them. Other earthquake-related risks which need to be considered by homeowners include the risk of tsunami damage or inundation in coastal areas, risks from nearby dams, and the risk from weak neighbouring buildings.

Earthquakes are, of course, not the only natural hazards risks which homeowners face, and other risks, such as those from flood, windstorms or wildfires, though perhaps not so devastating as earthquakes, may occur more frequently, and are expected to be more prevalent in the future. Any plan to protect the home against earthquakes should also take into consideration what could be done, at the same time, to protect from these other hazards.

In addition to knowing the hazards, it is also important for homeowners to understand the strengths and weaknesses of their building. The response of different building types to earthquakes is very different, as set out in Chapter 5. Generally unreinforced masonry buildings have a very poor earthquake performance, but confined and reinforced masonry perform better. For small buildings, well-braced timber-framed structures usually perform well. For taller buildings, steel-framed structures have historically performed better than reinforced concrete frame buildings, and buildings designed to current codes of practice should be safe. But particular features of a building can either enhance a building's performance, or worsen it. If a building is old and is not built according to current codes of practice, or is in poor condition, it is likely to perform worse; also if its shape is irregular, or it has been modified since it was built, this can affect its earthquake performance. Where possible, an experienced structural engineer should be consulted to help identify particular weaknesses, which could lead to earthquake damage.

Having identified the hazards and the vulnerability of the building, a plan for earthquake risk management needs to be made. The possibility of strengthening the building using retrofitting methods discussed in Chapter 5 should be considered. Some weaknesses, such as timber-framed buildings poorly bolted to foundations or unbraced cripple walls, can be relatively cheaply and easily eliminated. Others, such as additional shear walls in multi-storey buildings, the strengthening of masonry walls by the addition of reinforced jacketing layers, or the addition of separate steel frames are expensive and can cost up to 25% or more of the cost of a new building (Smyth et al. 2004; Spence et al. 2000).

Earthquake risk management should also include consideration of earthquake insurance. In some areas, a degree of earthquake insurance is compulsory or is required as a condition of a mortgage or bank loan. But cover under compulsory insurance schemes is generally limited, and it is possible to top up the cover from other insurers. The extent and cost of insurance needs to be carefully considered by comparison with the cost of strengthening to reduce the potential damage. Alternatively, renters can move to a different, lower-risk building. This will help raise awareness in landlords.

In apartment blocks, or buildings adjoined to neighbouring buildings in terraces or courtyard blocks, any retrofitting work will have to be agreed with all neighbours whose properties are affected. Where there is no single owner, this involves difficult decisions about the allocation of costs and agreeing a timeframe for vacating the properties for the work to be carried out. An effective residents' association will be needed to support retrofitting in such cases.

9.4.2   Securing Contents and Utilities

Homeowners (and to an extent tenants also) in earthquake areas need to give close attention to securing the contents of their home (including utilities), as contents damage can be a significant source of financial loss and falling contents can be a cause of injury. This is particularly true in homes which are already built to current standards of earthquake resistance. It is important to bear in mind that the design codes are primarily intended (see Chapter 7) to protect life by preventing collapse, and much damage and loss is possible in strong ground shaking, even when the building's structure itself experiences little damage. In recent moderate earthquakes in three European countries, damage to contents and non-structural items was found to constitute more than 60% of the total damage cost (Ferreira et al. 2020). And several studies (Horspool et al. 2020; Petal 2011) have found that a significant proportion of the non-fatal injuries caused by earthquake ground shaking in United States and New Zealand earthquakes were the result of being struck by unsecured contents.

Perhaps of greatest importance is the securing of utilities, particularly where piped gas is used as a fuel. Leaking gas lines can cause a fire to start long after the shaking stops, and this has been the cause of large-scale fires in several earthquakes in the USA as well as in the Mw6.9 Kobe earthquake in 1995 in Japan. Replacing rigid with flexible connections to appliances and installing excess-flow shutoff valves is effective in reducing the risk of leaks (Yanev and Thompson 2008). Damaged water mains can also cause extensive water damage, and electrical shorts can result in fires. All utility connections need to be inspected after an earthquake has occurred and should be shut off if there is evidence of damage. Unbraced water heaters are a major potential source of damage and should be securely tied to prevent toppling and rupturing pipework. Air-conditioning units also need to be securely anchored.

Heavy items of furniture, such as free-standing shelves, appliances and all large items like TVs need to be secured by wall straps attached to wall studs in timber-framed houses or wall anchors in masonry houses. And it is important that beds are not located where items on shelves or light fittings could fall on their occupants while asleep.

In addition to the building's contents, non-structural elements of the building such as plaster, partitions, false ceilings, chimneys, parapets and garden walls, can all be damaged or destroyed in an earthquake, and these can also be a cause of injury. It is worth seeking the advice of a structural engineer on how much risk these elements may pose, and how they can be better secured.

Detailed advice on securing contents suitable for local practice is available through national or municipal websites (http://www.fema.org in USA, http://www.eqc.govt.nz in New Zealand, http://www.metro.tokyo.lg.jp in Japan). These websites provide accessible advice which is likely to be useful elsewhere.

9.4.3   Preparing for an Earthquake

Each household should develop its own earthquake protection action plan, which should include the following elements:

- Assessment of the vulnerable components of the building, and a programme to address these as financial resources allow.
- A plan for securing the building's contents and utility connections as discussed above.

- Taking out insurance against earthquake damage to aid recovery and limit the financial hardship of an earthquake.
- Creating and maintaining a store of emergency supplies in the event either of an emergency evacuation or a period of time when normal supplies are unobtainable.
- Having an action plan for what to do during and after strong ground shaking occurs, which should include practicing 'drop, cover, hold' drills, learning first aid techniques to help the injured, knowing who to contact for assistance and having an evacuation plan in the event of serious earthquake damage or a tsunami warning, bearing in mind that aftershocks will always follow a major earthquake

Websites maintained by national or local civil defence organisations provide locally specific guidance on these topics (http://www.fema.org in USA, http://www.eqc.govt.nz in New Zealand, http://www.metro.tokyo.lg.jp in Japan).

The collaboration of local communities in the event of a natural disaster is often extraordinary (Solnit 2009), and developing an action plan for the local community can be a very important aspect of earthquake protection. A community action plan can involve specific roles for individuals including support for the authorities in damage assessments, house-to-house checks and contacting each household, first aid, firefighting, search and rescue, temporary shelter provision and maintaining emergency food, water and power provisions (Coburn and Spence 2002). Community groups can organise practice drills, with participation of schools, and organise local public awareness events. A well-organised local community can be the best basis, too, for mounting an advocacy campaign to seek government action for earthquake protection as discussed below.

9.4.4 Advocacy for Government Action

Ideally, citizens and the local community are where the initiative for action to improve earthquake protection should start. The most effective government actions for earthquake protection have come about through the campaigning of local community groups, supported by concerned professionals and political representatives at local and regional levels. Local advocacy potentially has an important role in setting the level of protection built into the building design codes, in land-use regulation, in the extent of enforcement of regulations and in the level of public expenditure allocated to building strengthening programmes, especially for schools, hospitals and critical buildings. Petitions, meetings with political representatives at the appropriate level of government, press campaigns and public meetings can all promote protection measures. Such campaigns often need to be maintained over a period of some years in order to be effective, taking advantage of political changes or disaster events, which can create a window of opportunity to get the desired changes implemented.

An example of a successful citizen-initiated advocacy is the story of Families for School Seismic Safety (FSSS), initiated in Vancouver, Canada, by a group of concerned parents led by Family Practitioner Dr Tracy Monk, summarised in Box 9.1, which led to the upgrading of several hundred at-risk schools in British Columbia over a 15-year period (Monk 2007). Tracy Monk attributes the long-term success of this campaign to four factors. First was establishing strong support from local scientists and engineers, both in identifying the levels of risk in existing schools, and also in proposing retrofitting solutions which could be adopted. Second was the adoption of what she calls a public health approach to the problem. Recognising that action was unlikely to be successful if the work had to be done within existing school budgets, FSSS campaigned at the State level for additional funding for life protection. Third was a lobbying campaign consisting of numerous meetings with politicians and public officials, and fourth was a very effective publicity and media campaign. Even after the State Premier had made a commitment in 2004 to spend $1.3bn on improving at-risk schools, continuing pressure was needed over another 15 years (so far) to keep the programme going.

The effectiveness of citizen's advocacy for school safety has been boosted in recent years through the World Bank's Global Program for Safer Schools, discussed in Chapter 7.

Box 9.1   Profile Dr Tracy Monk

Dr Tracy Monk: advocacy for school seismic safety

Tracy Monk is a Doctor and Family Practitioner in Vancouver, Canada. In 2002, she founded the advocacy group Families for School Seismic Safety (FSSS), which led to a commitment of US$1.3bn by the government of British Columbia for seismic upgrading of existing schools.

Tracy Monk

Asked what prompted her to start her campaign, she cited three catalysts. First, she is a parent, and her two daughters were of school age, 8 and 5 years old at the time; secondly, she had become vigilant about risks to her home when fire broke out in their garage, which could have destroyed their home and risked the children's lives; and thirdly, she was inspired by a media campaign mounted by two 16-year-old schoolboys at another school, who had discovered that school buildings in Vancouver were at disproportionate earthquake risk compared with other buildings, and asked why they were being required to put their lives at risk. 'I learnt more about advocacy from those two boys than from anywhere else', she said.

Her first action was to find out the facts. Through the School Board, she was able to trace a 1989 Vancouver School Board Report that showed that non-ductile reinforced concrete buildings, used for many schools in British Columbia, had a risk coefficient 10^5

(Continued)

Box 9.1 (Continued)

---

higher than wood-frame houses. Learning this 'made me physically ill', she recalled. She also talked with structural engineers including Carlos Ventura at the University of British Columbia and learnt about the possibility of retrofitting, and also discovered about the history of disproportionate damage to schools in earthquakes elsewhere (Chapter 7), including the USA. She asked 'how can we know all this and do nothing?' She found there was a reluctance among engineers to push the issue politically because of a concern about raising public anxiety, and because they might be accused of self-interest.

Most importantly, perhaps, as a Family Practitioner, she was able to view earthquake risk from a public health perspective. 'It was as though a vaccine is available, and no one is using it', she said, also arguing that the cost of the retrofit would be small compared with the cost of the lives saved, and crucially, that funding for life-safety should not be required to compete with the basic education budget for school books and teachers' salaries.

She put together Families for School Seismic Safety, a group of some hundreds of concerned parents, with active support from engineers and scientists and a number of concerned local politicians. The success of the FSSS advocacy campaign was, she believes, founded on four components (Monk 2007):

- An alliance with local scientists and engineers in developing and disseminating factual information outlining the problem
- A public health approach to the problem
- A lobbying campaign consisting of meetings with politicians and public officials
- A publicity and media campaign which created public support and focussed the politician's attention, creating a 'white knight' opportunity for government.

As a result of these activities over a period of more than two years, the then Premier of British Columbia in November 2004 made a $1.3bn commitment to seeing all schools brought up to life-safety standards by 2019.

The programme has not been without its challenges including how to balance architectural heritage preservation with seismic safety. The seismic mitigation programme was also delayed because of pending politically difficult decisions to close some schools because of falling school enrolment. And other buildings have been replaced rather than retrofitted. However, a 2020 report states that of 494 schools in the total projected programme, 174 had by then been completed, and a further 34 were under construction or proceeding to construction (http://www2.gov.bc.ca/assets/gov/education/administration/resource-management/capital-planning/seismic-mitigation/smp_online_report.pdf).

This is a very significant province-wide achievement, which has the potential to be replicated in other at-risk zones. Looking back on it Tracy Monk says that in some ways getting the government commitments was the easy bit. It was, she says, 'easier to get the money than to agree what should be done'.

Source: Modified from Monk (2007).

The Earthquake Engineering Research Institute in California (an international society, mainly of professionals in seismic safety related scientific and design disciplines) encourages its members to become effective citizen advocates on seismic safety policy in their own communities and has developed a toolkit to assist them in this effort (EERI 2019). It has developed policy positions on a number of aspects of achieving seismic safety worldwide, including promoting the use of confined masonry construction for emerging economies, eliminating the use of unreinforced masonry in school buildings by 2033, creating earthquake-resilient communities and promoting the adoption and enforcement of effective building codes with earthquake provisions. The toolkit provides advice on how to contact political representatives (legislators) at different levels and how to prepare for and conduct a phone-call or an in-person meeting.

A number of successful advocacy efforts on seismic safety in the USA were reviewed by Robert Olshansky (2005), on the basis of which he distilled a number of key pieces of advice. Above all was the conclusion that individuals can make a difference. But, to do so, they must

- Be persistent yet patient: recognising that it takes time to introduce the importance of seismic safety, both to the public and to decision-makers.
- Have a clear message: be able to explain in simple language what the problem is and how it can be solved.
- Understand the big picture: link seismic safety to other issues, such as safety of schoolchildren and the long-term sustainability of the local economy and look for ways to link seismic safety with safety from other hazards.
- Work with others: build networks of individuals through personal contacts as well as community organisations.
- Make seismic safety efforts permanent: establish an organisation, such as a seismic safety advisory committee which can maintain the necessary momentum over a long period.

Every one of the actions described in Olshansky's work owes its origin, or a boost in support, to an earthquake event, showing the importance of the post-earthquake window of opportunity to make changes. The cases he described were all from the United States, but the conclusions he draws are equally applicable to citizen advocacy for earthquake safety anywhere.

9.5 Scientists and Engineers

Science plays a key role in preventing disasters, preparing for earthquakes that cannot be prevented and to help communities recover from them. As scientist and engineers, we have a duty of care to our global neighbours to progress science and technology and disseminate the latest scientific findings and techniques to mitigate earthquake effects. Here are some examples of where professionals can make or have made a difference.

1) *Develop new materials and construction techniques to create safer buildings.* Cheaper ways of analysis, testing and manufacturing have paved the way for innovative designs in seismic isolation, vibration control and seismic resistant technologies. 3D printed

seismic resistant columns are no longer a figment of the imagination but becoming an affordable and practical reality. Artificial Intelligence (AI) and other remote methods have also enabled engineers to help policymakers and urban planners to survey and visualise strengthening programmes.

2) *Focus on culturally appropriate and innovative designs.* Safe shelter is a basic human right but as shown in the previous chapters, a lack of knowledge, skills and financial means have made safe housing an ambitious goal for some. We have witnessed first-hand how communities have struggled with building back, let alone building back better after earthquakes, as building materials are priced out of their means and they do not have the technical knowhow to construct for earthquakes. The Mw6.2 Yunnan earthquake in 2014 affected Ludian County in the Yunnan Province in China. In total, almost 81 000 homes collapsed, and 617 people died. In Guangming Village, out of a population of 61 families, 80% of the old buildings were local rammed-earth constructions, all of which collapsed. The villagers lost confidence in their traditional rammed-earth buildings and wanted to rebuild in brick or concrete. However, the price of construction materials rose sharply, and this option became unaffordable for most local villagers.

The poor thermal performance of the brick-concrete buildings and the lack of technical craftsmen also made it difficult for the villagers to rebuild a safe house. In response to the Mw6.2 Yunnan earthquake in 2014, the team led by Professor Edward Ng (Box 9.2) at the Chinese University of Hong Kong and Kunming University of Science and Technology set out to demonstrate how traditional methods can be improved to provide villagers safe, affordable, comfortable and sustainable homes. The team conducted research on the shortfalls of traditional rammed-earth buildings to formulate a new rammed-earth system. After repeated shake table and material tests, they designed and built a $148 \, m^2$ two-storey, new, rammed-earth house in four months, using locally sourced material and with the help of local workers. The prototype house won the World Architectural Festival award in 2017. The design was praised for its re-use of traditional material and construction methods, and its integration of new technology, combining ancient wisdom with modern know-how. The judges of the WAF were also impressed by the iterative research process which could be re-applied to anywhere in the world affected by seismic problems and low levels of wealth.

3) *Help with directing seismic research investments.* Given the limited resources for scientific research, international and national research councils have called upon the scientific community to help set priorities for investments. The involvement and active participation of scientists and engineers are vital in help set the agenda and future research in better earthquake resilience.

4) *Focusing on general disaster literacy rather than a single peril.* Given the infrequent and underrated nature of earthquakes, combining this natural threat with others, with the aim of improvising disaster literacy in general has proven to be effective. Studies have shown that integrating disaster reduction and poverty alleviation efforts for example, have had positive results in improving resilience. It is important to acknowledge the varied general disaster literacy of our target audiences around the world and find appropriate entry points to utilise limited resources.

5) *Effective risk communication* requires finessing the art of simple messaging. Public health messages that are vague with no associated actions will simply be ignored. Getting the science right is only half the battle. Dr Lucy Jones (Box 9.3) is a seismologist who has, after years of persistence, got the attention and respect of the local governments, media and the public.

Box 9.2   Profile: Edward Ng

Ng Yan Yung, Edward: innovative earthen architecture for rural China

Professor Edward Ng is an architect and Yao Ling Sun Professor of Architecture in the School of Architecture at the Chinese University of Hong Kong. He has been building bridges and eco-schools with local materials in the poor rural areas of the Loess Plateau in western China since 2003. His RIBA International award-winning design, the Maosi eco-school, completed in 2007, was his first attempt at using earth bricks in earthquake-prone areas in North-West China. Since then, raw earth directly from the ground has been the main construction material of his village projects. Since the Mw7.9 Wenchuan earthquake in 2008, and the Mw6.2 Yunnan earthquake in 2014, he and his team have developed improved techniques and worked in impoverished villages to rebuild houses and community facilities together with villagers.

Edward Ng on site in Sichuan, 2009

Earthen construction has a long history in China. Historically, earth has been the main building material for traditional village buildings as it is inexpensive and readily accessible, and it has good thermal mass. Until recently in China, many earthen buildings were still being self-built by villagers using traditional methods.

In China, over 40% of the population, or some 564 million people, are still living in rural areas – many of them are poor with a monthly income of less than ¥1000 RMB (or $150 USD). As China continues to modernise, rural construction and village development has become a key socio-economic-political challenge for the government. The local officials face several problems. First, during the recent and ongoing urbanisation process, many villagers leave their homes and become migrant workers in the urban areas, causing a lack of a skilled labour force in the villages. The traditional earthen techniques are no longer continued or developed, resulting in earthen structures of poorer quality being erected. Second, as migrant workers bring back news and images from the city, rural villagers have now come to view these traditional earthen buildings as symbols of poverty and backwardness because of their low quality and design, with

---

(Continued)

Box 9.2 (Continued)

poor indoor environmental quality (daylight and ventilation). Third, traditional earthen houses are not earthquake-resistant, and in earthquake prone areas, they are considered dangerous. Since 2011, the central government via its 'beautiful countryside construction' movement to revitalise the Chinese countryside has come to recognise the values of traditional earthen houses and has requested local officials to observe a checklist of mandates. This has created an opportunity for Ng and his team to expand their work.

Ng believes that by re-examining and modernising traditional earthen techniques it is possible to mitigate the poor and unsafe conditions of villagers living in traditional earthen houses while at the same time re-affirming their ancestral heritage and conserving their cultural identity. 'Giving the villagers just a house is not enough', Prof Ng explained, 'You must give them a better life after an earthquake, showing them interventions that they can build themselves that are better environmentally and stronger structurally'.

Ng and his team's work is based on the three L's – local technology, local materials and local labour, giving the villagers full control. The designs adopt a 'High-Science-Low-Technology' approach where state-of-the-art modelling and testing are carried out to optimise the use of materials and construction techniques. Social surveys ensure the client's needs are met. Once on site, the focus is on low technology, allowing a farmer-villager and his family to self-build. New tools and methods have also been introduced to help with the construction process.

In the past, building houses was a village's main social event. The whole village would come together helping each other with construction, and a village feast always followed. 'That way, the house belongs to the village. We need to revitalise this tradition', explains Ng. The building projects go beyond a single dwelling, knowledge is transferred.

To date, more than a hundred villagers including some 30 female villagers have been trained and more than a hundred new houses have been built on site in China. The first village house constructed by a fully female team has also been completed. A new earthen building code is being developed by the provincial government of Yunnan and a new training college has been established in Kunming, Yunnan.

'Our work has just begun!'

Traditional rammed earth building destroyed by the Mw6.2 Yunnan earthquake in 2014

The post-earthquake reconstruction demonstration project of Guangming Village, Yunnan, China which won the World Building of the Year Award at the 2017 World Architecture Festival (WAF) in Berlin

Box 9.3   Profile: Lucy Jones

Lucy Jones: raising public earthquake awareness

Dr Lucy Jones during an interview after a 6.1-magnitude quake struck near Joshua Tree in 1992 (Credit NBC Los Angeles)

For 33 years, Dr Lucile (Lucy) Jones was the face of the US Geological Survey, a seismologist Southern Californians trust to give them the facts at unsettling times. A recent

(Continued)

Box 9.3   (Continued)

article in the Los Angeles times read 'Who doesn't trust Lucy Jones in a crisis?' This is no small accolade. Lucy Jones has over the years become the entrusted voice. She has a unique gift in communication. The ability to explain complex science to the public in ways that helped them understand. 'When we got scared, she talked us down without ever talking down to us' (Lelyveld 2020). She combines her knowledge of science with enthusiasm, reason and gentle persuasion.

Dr Jones' first earthquake response interview was in 1985, but one of her most memorable appearances was in 1992 when she appeared on live television with her son Niels, who was one at the time, sleeping in her arms. 'I'm everybody's mother', she likes to joke, aware that her gender – while not an asset when she was at MIT in the 1970s – is now a plus.

Many articles have been written about this national treasure who challenged the social norm in the 1960s and 1970s to study and excel in science at school and university. She studied physics and Chinese at Brown and graduated with a B.A. in Chinese language and literature. Having studied geophysics in her senior year, Jones went to MIT to get a doctorate in geophysics – one of just two women at the school pursuing an advanced degree in that subject, examining foreshocks. In 1979, while still at graduate school carrying out research on the Mw7.0 Haicheng earthquake in 1975, Jones was one of the first Westerners to visit China as a Fulbright Fellow, spending a total of 12 months between 1979 and 1983 conducting research on Chinese earthquakes with colleagues at the Chinese Earthquake Authority in Beijing.

During her time at the USGS, she worked as a researcher in seismology, developing methodologies for assessing the short-term changes in the earthquake hazard during earthquake events that have been used repeatedly in California to advise the government and the public. For eight years, she served as the Scientist-in-charge of the Pasadena office of the USGS. Jones helped create the California Integrated Seismic Network and has played a key role in improving US's responses to natural disasters. Her research is the basis for all earthquake advisories issued in California. However, in the latter part of her tenure at the USGS, her focus turned to policies, working with policy-makers to empower them to make better use of science and to appreciate the necessary trade-offs.

In 2014, Jones was seconded to the Los Angeles City Hall and served as the Science Advisor for Seismic Safety to Mayor Eric Garcetti. Through her determination and persistence, she secured a retrofitting plan that would upgrade over 15 000 older, substandard structures, largely built of non-ductile concrete, in the city.

Amongst her many achievements at the USGS, one of the most remarkable is the Great California ShakeOut. A comprehensive report based on a Mw7.8 earthquake scenario along the southernmost San Andreas fault which became the basis of an annual earthquake drill. In 2019, 11 years after its inception, more than 60 million people participated and practised 'drop, cover and hold on' and other earthquake preparedness actions around the World.

Jones retired from the USGS in 2016 and in honour of her work, 30th March was declared 'Lucy Jones Day' in Los Angeles that year. But she was not done. Instead of sitting back and enjoying her well-earned retirement from public service, Jones has set up a centre dedicated to bridging science and public policy, examining other risks such as climate change and epidemics the world must learn to adapt to today. She also teaches a class at Caltech entitled 'Science Activation' on how to get science used. Her motto, 'What good is scientific knowledge if people don't use it?'

Dr Lucy Jones during the Great Shakeout earthquake drill of 2012 with Mayor Villaraigosa of Los Angeles

9.6 NGOs

At grassroot levels, the involvement of international and national non-governmental Organisations (NGOs) has been instrumental in bringing about much needed investments and skills to areas prone to earthquakes.

International and local NGOs forming partnerships to ensure a sustainable future for earthquake-prone countries is not a new concept but one that is hard to maintain, especially as projects are reliant on funding and personnel. The personnel issue is often not a lack of commitment from the participants but the transient nature of the sector. In 2007, the authors visited Pisco in Peru after the earthquake that year and found that locals affected by the event were desperate for knowledge and how to build back better. They directed the authors to a reinforced adobe house that was built by the Japanese NGO JICA, and showed us how it had withstood the earthquake with no damage while other adobe houses around it collapsed. However, there were no legacies from this project, knowledge that was relayed back to the local people, or at least those that remained in the village. This lost window of opportunity is a stark reminder that in order to make a lasting impact, the sustainability and longevity of the projects and must be considered.

In the past, the types of work NGOs have been involved with in earthquake risk reduction include both pre- and post-disaster programmes. The most recognisable form of

NGO intervention comes in the form of post disaster humanitarian aid. To help with maintaining an international standard, the International Federation of Red Cross and Red Crescent Societies developed a code of conduct for humanitarian aid and lists the names of all international and local NGO signatories of the code on their website. Depending on the location of the earthquake, the participants in aid will differ, however in recent years a few of the main groups have been MSF (Médecins Sans Frontières), OXFAM and Save the Children.

The main issue with NGOs is the timing and duration of their involvement. They provide much-needed emergency resources, skills and facilities at a time of acute need, but typically, they demobilise after one to three months. There are sometimes blurred boundaries between when a NGO's responsibility ceases and when others (and who) take over. This is especially problematic when it comes to housing. Emergency tents are usually set up in available open spaces post-disaster, but these are often the exact spaces which could be available for reconstruction and rehousing the displaced. The issue of who has the responsibility of setting up transitional housing is also hotly debated. In the case of the Mw7.0 Haiti earthquake in 2010, millions of dollars were spent on transitional shelters for very poor households that cannot afford to upgrade them. The sheer scale of damage and number of people left homeless triggered an outpouring of aid, and a plethora of agencies flew into Haiti. However, this aid from large and small NGOs was often uncoordinated and micro-projects searching for ideal housing solutions created pockets of excellence and disparity amongst communities and recipients. Hundreds of thousands of people rebuilt their homes quickly, either in the same vulnerable sites or in new sites, exacerbating the capital city Port-au-Prince's sprawl. Without technical assistance, unsafe building practices were repeated, and sadly vulnerability was built back in.

The work of NGOs in Haiti did provide much-needed safe housing and basic services to devastated communities. However, many NGO-led projects were carried out in isolation, with little thought of how they connected physically, socially or economically to other neighbourhoods, or to the wider city. The lack of local authority and community involvement was identified as a key oversight in the efforts of reconstruction and recovery in Haiti (Gill et al. 2020).

Two outfits that have worked tirelessly in local capacity building and transferring earthquake knowledge are GeoHazards International (GHI, https://www.geohaz.org/) and Build Change (https://buildchange.org/). GHI is a non-profit focused on reducing preventable death and suffering from natural hazards in the world's most vulnerable communities. Focused on working with local partners and NGOs, they deliver material, whether in writing or with tools, to improve earthquake knowledge and understanding of risks. One such projects was the introduction of earthquake safe desks in Bhutan. This simple but effective intervention does not replace retrofit programmes but does provide the necessary time for upgrade. In total, 150 schools were damaged after moderate earthquakes in 2009 and 2011 in the Kingdom. The Mw7.8 Gorkha earthquake in 2015 was a harsh reminder of what could happen to the unreinforced masonry schools in Bhutan. In 2016, GHI partnered with the ministry of education and, with funding from catastrophe modelling company AIR and various agencies, sent two earthquake desk designers to train five furniture manufacturers to make the desks in Bhutan. The week-long programme ended with a live demonstration to the different government ministries, schools, the armed forces and local media. A simple experiment where building rubble was dropped at height onto a standard school desk and onto the earthquake safe desk was shown (Figure 9.6).

Build Change was founded by engineer Elizabeth Hausler in 2004 and has been a member of the Clinton Global Initiative since 2010. Its ethos is one driven by Hausler's personal experiences in India following the Mw7.7 Bhuj earthquake in 2001. She envisions a Theory of Change, for ensuring vulnerable communities globally are empowered to build permanent safe housing in the wake of disasters.

Their most ambitious programme to date is the '10 in 10' initiative. Launched in 2014, its ambition is to empower 10 million people in emerging nations to live and learn in safer homes and schools by 2024. The daughter of a mason, Hausler has instilled the need for hands-on training in her staff, and they teach local people how to build, making small changes to existing local architecture to strengthen housing against disasters. In doing so,

Figure 9.6 Photographs taken of the live demonstrations of the earthquake desk. The top shows the drop of a 356 kg load crushing an ordinary school desk; the bottom photo shows the drop of 422 kg load and the desk surviving the drop and retaining a safe space underneath.
Source: Credit: GHI

knowledge is transferred, capacities are built with small financial incentives. There are four pillars to their approach:

1) Power to the households and to women, who often head the households in terms of everyday decisions or spend the most time in the dwellings.
2) Provide conditional financial cash grants to improve building standards.
3) Prevent the disasters in the first place. The cost of retrofitting is 20–30% of the cost of rebuilding after the events. Build Change believes it is worth investing US$30k now rather than the US$200k to rebuild.
4) Use technology to help visualise the impact of change to building owners and decision-makers. Recent technologies have reduced the time taken to review retrofit options by 97% using Virtual Reality. Build Change have also used Artificial Intelligence (AI) to assess entire neighbourhoods to compare strengthening buildings versus doing nothing.

9.7 Insurers

One way in which mitigation activity can be promoted is through the development and spread of natural hazards insurance. Although the primary purpose of insurance schemes is to spread the burden of paying for post-disaster recovery to a wider group, nationally and internationally, they create significant opportunities for mitigation. It is very much in the insurer's interest to limit the losses. One way to achieve this is to use differential premium rates and different deductibles or coinsurance according to the extent of mitigation activity carried out by the policyholder. Since the early days of fire and marine insurance, such mitigation incentives have been used, and they are common today in household insurance policies. Insurers in most earthquake countries already offer lower rates for types of building which have proved less vulnerable in earthquakes. This encourages good building practice for new building. Likewise, sensible location of new building can be encouraged by offering different rates for lower-risk sites. Given the experience of numerous earthquakes over 25 years, Mexico City has one of the most sophisticated rating structures, identifying three different construction types and six separate risk zones within the City. Over the last two decades, these differential rates, varying by a factor of 10, have contributed to a substantial shift in the commercial centre of the city from high-risk to lower-risk geographic areas.

However, differentials in most earthquake insurance schemes do not accurately reflect relative risks (Coburn and Spence 2002; Woo 1999). Indeed, because of the heavy geographical concentration of losses from natural disasters, many insurers are reluctant to offer insurance, or will only offer it at unacceptably high premium rates. Thus, commercial insurance coverage in some areas of high risk is patchy, and a large part of the risk remains uninsured. For example, in the largest US catastrophes of recent decades, insurance coverage of the losses has been no more than approximately 50% of the total. In Japan, only 3% of the losses in recent earthquakes were covered by insurance. Most of the low-income countries also have a very low rate of insurance penetration. Table 9.1 shows, based on EM-DAT data derived from Munich Re, the extent of the loss in the most damaging recent earthquakes which was covered by insurance.

Table 9.1 Most costly events since 2000, and proportion of insurance cover.

Date	Country	Event name	Magnitude (Mw)	Total loss (US$bn)	Insured loss (US$bn)	Insured (%)
11/03/2011	Japan	Tohoku	9.1	210	37.5	18
12/05/2008	China	Wenchuan	7.9	85	0.3	0
27/02/2010	Chile	Maule	8.8	30	8	27
23/10/2004	Japan	Niigata	6.6	28	0.76	3
16/04/2016	Japan	Kumamoto	7.0	20	5	25
20/05/2012	Italy	Emilia-Romagna	6.0	15.8	1.3	8
22/02/2011	New Zealand	Christchurch	6.1	15	12	80
16/07/2007	Japan	Niigata	6.6	12.5	0.335	3
12/01/2010	Haiti	Haiti	7.0	8	0.2	3
20/04/2013	China	Lushan	6.6	6.8	0.023	0
04/09/2010	New Zealand	Darfield	7.0	6.5	5	77
19/09/2017	Mexico	Puebla	7.1	6	2	33
08/10/2005	Pakistan	Kashmir	7.6	5.2	0	0
25/04/2015	Nepal	Gorkha	7.8	5.2	0.1	2
21/05/2003	Algeria	Boumerdes	6.8	5	0	0
03/08/2014	China	Yunnan	6.2	5	0	0
24/08/2016	Italy	Amatrice	6.2	5	0.075	2
26/12/2004	Indonesia	Indian Ocean tsunami	9.1	4.5	0.225	5
14/11/2016	New Zealand	Kaikoura	7.8	3.9	2.1	54

Source: Based on EM-DAT data derived from Munich Re.

The differences in insurance coverage are startling. New Zealand, with 80% and 77% cover for the Darfield and Christchurch earthquakes, has by far the most extensive cover, owing to the success of its national earthquake insurance scheme, EQC, described below. Japan, Mexico and Chile recorded cover from 18 to 33%, though much lower in Japan's Niigata earthquakes in 2004 and 2007. In the lower-income countries, India, Indonesia, Pakistan and Nepal, insurance cover was negligible. In some middle and high-income countries (China, Italy), it was also very low because of the expectation that building damage costs will be paid by the government, either in reconstruction projects, or through grants.

These figures can be compared with insurance cover for European windstorm, exceeding 90% of probable losses. Reasons for low coverage of earthquake insurance in the richer countries, including the USA, for which no earthquake appears in Table 9.1, may include the following:

- A few insurance companies not offering, or not promoting, such insurance.
- Relatively high levels of premiums putting off homeowners.
- Banks and loan agencies not requiring it.
- The relatively low risk of damage exceeding the (typically 15%) deductible.
- The assumption that government aid will assist those who are not insured.

In Japan, by law, the maximum claim for any one event is limited to a value far less than replacement cost. A further disincentive is that, as noted above, the differences in rates offered by insurers between different zones are often far less than the actual relative risks; and the same is true for different building types. Thus, the lower risks subsidise the higher, reducing the attractiveness of insurance to the majority.

Due to the perceived benefits of natural catastrophe insurance, and the problems of ensuring sufficient uptake of insurance offered through normal commercial methods, the governments of a number of countries (or states) have chosen to intervene into the purely commercial insurance market, either creating their own insurance pool, or regulating the terms or scope of commercially available insurance. These schemes are interesting, offering models which can be studied by other countries for the opportunities they provide to create incentives for mitigation activity. Three different existing natural disaster insurance schemes which have been in existence for some time are briefly examined.

9.7.1 New Zealand Earthquake Commission

In 1941, the government set up a special war damage insurance pool funded by a levy of 0.25% of insured value on all fire premiums. This was then extended to cover earthquake, establishing the pool as the Earthquake and War Damage Fund. After the war, the levy was reduced to 0.05%.

In 1993, the scheme was restructured as the Earthquake Commission (EQC), and the fund renamed as the Natural Disaster Fund. The government guarantees that this fund will meet all its obligations. EQC buys a substantial amount of reinsurance each year to enhance its ability to pay claims and to minimize the possibility of calling on the government guarantee.

Earthquake Commission cover is compulsory for residential property, and private insurance companies have responsibility of collecting the premium. All residential property owners who buy fire insurance from private insurance companies automatically acquire EQC cover. If the dwelling or personal possessions are more valuable than the maximum amounts EQC will cover (in 2020 set at $150 000), and homeowners can arrange extra cover with a private insurance company. EQC cover insures dwellings, personal property and adjacent land and structures. The cost of EQC cover is fixed and charged in the form of a proportion of total sum insured, rather than considering factors like location and building type.

EQC worked well until 2010, because earthquake events until then had typically led to relatively small numbers of claims. But EQC was overwhelmed by the huge number of claims (over 460 000) in the Canterbury earthquakes of 2010/2011, particularly when it was required by the government not just to make cash payments, but to manage a mass-scale repair programme. Many claimants considered their claims had been mishandled, and there were many disputes. A new EQC law was enacted in 2019 with the intention of improving claims management, which also increased the payment cap, while removing contents cover. And a public inquiry, which reported in 2020 (Government of New Zealand 2020) has recommended many changes in EQCs claims management and communication procedures.

However, in spite of these failings, EQC (with over 98% of residential buildings covered) has been very successful, along with effective building regulations, in ensuring that buildings in New Zealand are generally soundly built, and this was an important part of the reason why the number of building collapses in the earthquakes of 2011 and 2016 was comparatively low.

9.7.2 Turkish Catastrophe Insurance Pool (TCIP)

In Turkey, following the disastrous earthquakes of 1999, the government of Turkey, supported by the World Bank, developed a compulsory national insurance scheme. Through a 1999 decree (a law since 2012), all existing and future privately owned property was required to contribute to the Turkish Catastrophe Insurance Pool (TCIP). Non-engineered rural housing and fully commercial buildings are excluded. The intention was to create a fund by homeowners' annual payments for use in disasters so that no one would be left homeless, with a nominal sum (capped at about US$50 000) being disbursed immediately to homeowners who are left homeless (Gülkan 2019).

The model of the pool management was patterned after New Zealand's Earthquake Commission (EQC), and the California Earthquake Authority (CEA). However, TCIP is different from either of these public institutions, in that management is carried out by a commercial insurer. The risk is reinsured internationally by reinsurance policies supported by catastrophe modelling (Bommer et al. 2002). Most of the insurance companies in Turkey sell TCIP policies. In 2019, TCIP had a national market penetration of 50% with some nine million policies sold. The premium paid differs according to the earthquake zone and the form of construction, with higher premiums for masonry buildings than for reinforced concrete structures, although there is no difference according to the standard of construction.

TCIP has conducted many public education and awareness-raising campaigns. The extent of coverage is increased by regulations that require TCIP coverage if properties are to be allowed connections to utilities, or to be sold. An important feature of the original law was the denial of state assistance to homeowners who have not purchased insurance.

However, in a number of earthquakes, non-insured families have also received compensation (Gülkan 2019), reducing the incentive to purchase insurance.

Though not perfect, TCIP, in its 20 years of existence, has done much to improve earthquake awareness and stimulate better building practices in Turkey. Following the example of Turkey, the World Bank has supported the implementation of several similar schemes, for instance in Romania. National natural catastrophe insurance pools are currently active in 12 countries worldwide (Franco 2014).

9.7.3   California Earthquake Authority (CEA)

The California Earthquake Authority came into existence following the 1994 Northridge Earthquake in Southern California, in which the existing insurance industry was so badly hit that most insurers stopped offering earthquake insurance. CEA was set up in 1995 by State legislation to offer affordable insurance to all householders, and the coverage of residential insurance has steadily grown since then and now covers over 1 million homes (CEA 2020). In addition to offering insurance, CEA very actively promotes earthquake risk mitigation and supports action by householders to retrofit older properties through its 'brace and bolt' programme, which by 2020 had reached 11 000 pre-1980 properties.

Earthquake insurance is not compulsory in California, and a significant proportion of California residents do not carry insurance for the reasons stated above, but particularly because they consider their homes strong enough to avoid any damage exceeding the deducible. But by active support to earthquake risk mitigation CEA has contributed to the steady reduction of earthquake risk in California, as noted in Chapter 8.

9.7.4   Micro-insurance

Micro-insurance has been widely proposed (Shah 2012) as a means to bring the benefits of insurance cover to the many millions of poor families, in urban and rural areas, who cannot afford conventional insurance. The idea is that, for a small annual premium, attached to other forms of community financing, and paid by enough families, a pool could be created at a national or regional level sufficient to pay for repair or reconstruction of houses damaged in a large earthquake. For a scheme in rural China, modelling by risk modelling company RMS (2009) indicated that an annual premium of just $1.50, paid by 55 million households, would be sufficient to pay for the expected annual earthquake losses, and the costs of administering the scheme. In the Indian State of Gujarat, a scheme has been proposed that would insure local micro-finance institutions, so that in the event of an earthquake, existing loans of those whose houses were damaged would be written off.

However, the effort and ingenuity put into the planning of such schemes has not yet translated into much effective action, and a study by Beijing Normal University identified five primary reasons for a lack of trust in insurance by poor families (Shah 2012):

- Lack of understanding of insurance
- Do not trust the insurance companies to pay
- Can't afford premium
- Don't think any major disaster will occur
- Government relief makes insurance unnecessary

It has also been argued by insurance expert Guillermo Franco (Franco 2014), that cash payments are not necessarily the best way to assist the poorest disaster victims, who need other forms of technical and community support to help them build future resilience as discussed in Chapter 7.

9.8 The Way Forward

UNDRR published a report on 'Strategic approach to capacity development for implementation of the Sendai Framework for Disaster Risk Reduction: a vision of risk-informed sustainable development by 2030' in 2018. This 111-page report includes a list of roles and responsibilities for different stakeholders in disaster risk reduction (DRR). The stakeholders include the UN, International Organisations and International Financial Institutions; Regional Organisations including Intergovernmental Organisations; National, Local and Sub-national Governments; Private Sector and Professional Organisations; NGOs and CSOs (civil society organisations); Education and Research Institutions; Individuals and Households and the Media. The document provides generalised advice on the capacity development roles and responsibilities of these stakeholders and gives high-level guidance in six critical areas of identified need.

1) Developing and Strengthening DRR Fundamentals
2) Institutionalising DRR Capacity Sharing
3) Using Risk Information Before and After Disasters
4) Establishing Collaborative Action for DRR at the National and Local Levels
5) Strengthening External Support Mechanisms
6) Advancing and Expanding DRR Capabilities

As we strive for an equitable DRR future, it is also important that we understand and implement DRR measures that are sensitive to the cultural, social and financial constraints of earthquake-prone countries. Local capacity development is at the heart of the DRR and resilience agenda.

In this book, we have relayed our knowledge of earthquakes and their consequences, provided accounts of our own experiences of visiting areas affected by earthquakes in the past and summarised our peers' collective views of the physical vulnerability of buildings to earthquakes around the world. We have shown through examples of bad and good practices, the challenges but also the windows of opportunities we, the international community of DRR stakeholders, must face to curb the trend of earthquakes becoming disasters, especially in developing countries.

The message is simple. With affordable protection actions, buildings collapsing and people dying from earthquakes are largely preventable. Much like vaccines to infectious diseases, the consequences of earthquakes can be eradicated, but it takes commitment from all the stakeholders involved, not just the international funders and governments, but also from those who will be directly affected.

It is evident from our review in this book that although there has been progress made in earthquake risk reduction, much remains to be done. The areas where we can make the most difference can be broken down into three main categories.

9.8.1 Understanding Risk

We need to continue to advance our understanding of the underlying threat through research in all areas of earthquake science. Investing in science and technology to help with identifying, quantifying and monitoring the hazard will help us mitigate the impact of earthquakes. Research also has a part to play in developing innovative solutions at low costs.

One area that is underdeveloped in our understanding of risk is in quantifying exposure. Globalisation and rapid urbanisation have changed not only the physical landscape of most countries but have also introduced new working and living patterns. The recent Covid-19 pandemic has highlighted the fragilities and the interdependent nature of the modern society. The housing of the most socially vulnerable in informal slums have been identified as physically vulnerable in earthquake risk reduction programmes for some time, but over-crowded workplaces (such as factories and food-processing plants) are not recognised as particularly vulnerable buildings. In some countries, millions of people work for long hours in proximity in substandard buildings and conditions. The non-residential sector is disparate and is not yet collectively mapped nor well understood. Multiple and parallel failures of these densely populated buildings would significantly burden search and rescue and emergency services, and result in short- and long-term social and economic consequences as well. Investment in quantifying assets and risk in these sectors is urgently needed.

9.8.2 Communicating Risk

Our efforts in understanding risk are null and void unless we can effectively communicate them to different audiences to trigger action. The widespread use of social media and the ability to access information across the globe 24 hours a day provide opportunities to learn and share. In terms of education, our reach is now far wider than before, and the organisations such as the Global Program for Safer Schools and the Aga Khan Foundation have taken advantage of this to provide online learning platforms, sharing experiences and knowledge across the globe.

Others have explored novel ways of communication, for example through art. Figure 9.7 shows the centrepiece at the World Bank's Understanding Risk Finance Pacific forum in 2018. 'Why not transform the data behind these risks into something people absorb with their senses, that grabs them in an emotional way?' the artist Suarez asked. The sculpture depicted a large set of datapoints and their interrelationships, telling multiple intertwining stories about people, nature and finance.

For the attendees at the conference, the government officials, development institutions and humanitarian relief organisations, the sculpture became a natural talking point and was a visual reminder of the importance of developing financing solutions to manage the devastating impacts of disasters. In countries where they have had recent devastating earthquakes, museums have been built to remind future generations of the threat and the great consequences, e.g. the Kobe Earthquake Museum in Japan and Quake City at the Canterbury Museum in Christchurch, New Zealand.

Messages must be precise and actionable and catered for the right audience. News media can be an ally and a curse. The late Sutopo Nugroho from the Indonesian National Disaster

Figure 9.7 Pablo Suarez explaining 'Risk and Time: A Data Sculpture on Nature, Disasters, and Finance'. *Source:* Photo Credit World Bank – Nausicaa Favart Amouroux, Fondation Suzanne Bastien, Vanuatu.

Agency BPBD was someone who saw the benefits of having journalists as his allies to get accurate information out. Working closely with journalists, he gained respect from the media and the public, always available for press releases, answer phone calls and respond to interviews. His tone was never patronising but one which educated the audiences, very much like Dr Lucy Jones (Box 9.3), he became the scientist people trusted. Initiatives to educate and communicate risk to the media are now part of the Global Facility for Disaster Risk Reduction's (GFDRR) programme.

9.8.3 Incentivising Risk Reduction

How can we incentivise earthquake risk reduction and in what form? As highlighted in the book, the infrequency of earthquakes has meant that implementing mitigation measures has not been pressing on the political nor personal agendas of populations at risk. To make a difference, we must change tack.

Are we using the wrong metric to quantify the earthquakes? Fatalities are reducing in many countries, but costs of earthquakes are rising dramatically, and earthquakes will continue to be among the costliest single disaster events in the future e.g. Tohoku, Christchurch. Can we gain more attention by using a different metric, like economics, to communicate the severity of earthquake risk to the public? The cost of doing nothing in terms of earthquake mitigation far outweighs the cost for preparedness, 'an ounce of prevention is worth a pound of cure'. There is a perception that earthquake resistance is very expensive.

When we think of earthquake-resistant buildings, high technological solutions like base isolation comes to mind. People are often surprised to find that the cost of making a building safer is not high. 'Building better' is a knowledge-driven process, potentially requiring higher skills which would require better trained labour and more expenditure on preparation and design time, but in general requiring little more in materials or increased construction time.

However, building safely may not be an option for some of the most socially vulnerable. For families with low disposable incomes, spending on upgrading a house is a luxury, not a necessity. Promoting measures for earthquake protection alone, however worthy the cause, will not be top priority. Understanding the constraints of the affected population and working with the communities to understand their needs are vital in these circumstances.

Financial incentives through the issue of grants or lower insurance premiums have been used in the past. However, given the uptake has been patchy (see Section 9.7) perhaps incentives should also be considered in non-monetary forms. The worldwide movement against use of plastics maybe provides a useful analogy. This movement is one that has crossed both borders and traditional political divides. Plastics have been a way of life for generations, providing a cheap and light alternative to other materials. The unwarranted use of plastic microbeads in beauty products became the tipping point in 2015 and the plastic movement has since snowballed. The reduction in supermarket shopping bags and single-use cups is only the tip of the iceberg, but how the public's attitude to this material has changed in such a short time has defied expectation and is very encouraging. The movement to build better in earthquake areas could learn from this experience.

The climate change agenda and the urgency to act have been prominent in recent years. The public have become more involved, partly helped by social media platforms. The younger generations believe that individual actions count. The need to change the way we live and reduce the use of valuable resources will prompt changes to the way we build. This gives us a window of opportunity to provide holistic solutions to our clients, considering their needs, and introducing feasible measures that will protect lives under multiple natural threats.

Why do buildings collapse in earthquakes? The reason is not one just of poor construction but a combination of a lack of understanding of the underlying threat, knowledge of how to build appropriately and awareness that action can make a difference. Through our combined years of research, we believe earthquake risk reduction fundamentally comes down to individuals who are driven by their belief in improving earthquake safety. It is through the *information providers* that we can improve our understanding of earthquakes and through their platforms that we are able to share this knowledge. We need *advocates* working with governments and international agencies to push the agenda for risk mitigation policies and regulations at international/national/local levels. Crucially, at a local level, we need the *implementers*, like all our profiled game changers who find novel ways of engaging with people affected by earthquakes to make the necessary changes to their lives, whether physical or behavioural.

As our world population increases and more people are congregating in areas of risk, we have a duty of care as professionals to communicate earthquake risks, provide feasible solutions and help people engage constructively with reducing earthquake risks worldwide.

References

Bommer, J., Spence, R., Tabuchi, S. et al. (2002). Development of an earthquake loss model for Turkish catastrophe insurance. *Journal of Seismology* 6: 431–446.

Cattanach, A. (2018). 12 projects over 12 years: reflections from implementing low damage designs. *Presented at the 17th US–Japan–New Zealand Workshop on the Improvement of Structural Engineering and Resilience* (12–14 November 2018). Queenstown, New Zealand.

CEA (2020). *History of the California Earthquake Authority*. CEA.

Coburn, A. and Spence, R. (2002). *Earthquake Protection*, 2e. Wiley.

EERI (2019). EERI citizen advocate toolkit. https://www.eeri.org/wp-content/uploads/2020/11/EERI-Citizen-Advocate-Toolkit-Nov-02-2020.pdf (accessed December 2019).

Erdin, B. (2018). Veli Göçer returned to construction business. https://www.sozcu.com.tr/2018/gundem/veli-gocer-insaat-isine-geri-dondu-2580183/ (in Turkish) (accessed 15 June 2020).

FEMA (2020). Business quakesmart toolkit. https://www.ready.gov/sites/default/files/2020-04/ready-buisiness_quakesmart_toolkit.pdf (accessed 10 May 2020).

Ferreira, M., Meroni, F., Azzaro, R. et al. (2020). What scientific information on non structural elements seismic risk do people need to know? Part 1: compiling an inventory on damage to non-structural elements. *Annals of Geophysics* 63: AC04.

Franco, G. (2014). *Earthquake Mitigation Strategies Through Insurance*. Encyclopedia of Earthquake Engineering.

Gill, D., Smits, L., and Stephenson, M. (2020). *Learning from Community Planning Following the 2010 Haiti Earthquake*. London: IIED http://pubs.iied.org/10857IIED.

Government of New Zealand (2020). *Report of the Public Inquiry into the Earthquake Commission*. Wellington: Government of New Zealand.

Gülkan, P. (2019). Survey response.

Horspool, N., Elwood, K., Johnston, D. et al. (2020). Factors influencing casualty risk in the 14th November 2016 Mw7.8 Kaikōura earthquake, New Zealand. *International Journal of Disaster Risk Reduction* 51: 101917.

Lelyveld, N. (2020). *Who Doesn't Trust Lucy Jones in a Crisis? Here's What She Has to Say About the Coronavirus Pandemic*. LA Times. https://www.latimes.com/california/story/2020-06-27/who-doesnt-trust-dr-lucy-jones-in-a-crisis-heres-what-she-has-to-say-about-the-coronavirus-pandemic.

Monk, T. (2007). School seismic safety in British Columbia: a grassroots success. https://www.crhnet.ca/sites/default/files/library/Monk.pdf (accessed 6 April 2020).

OECD (2020). OECD toolkit for risk governance: good practices. https://www.oecd.org/governance/toolkit-on-risk-governance/goodpractices/?hf=10&b=0&sl=trig&s=desc (accessed 6 October 2020).

Olshansky, R. (2005). Making a difference: stories of successful seismic safety advocates. *Earthquake Spectra* 21: 441–464.

Petal, M. (2011). Earthquake casualties research and public education. Chapter 2 in human casualties in earthquakes. In: *Human Casualties in Earthquakes* (eds. R. Spence, E. So and C. Scawthorn). Springer.

Pfefferbaum, B., Pfefferbaum, R.L., and Van Horn, R.L. (2018). Involving children in disaster risk reduction: the importance of participation. *European Journal of Psychotraumatology* 9 (sup2): 1425577. https://doi.org/10.1080/20008198.2018.1425577.

RMS (2009). RMS completes micro-insurance quake project for rural China. *Insurance Journal*. https://www.insurancejournal.com/news/international/2009/06/08/101160.htm (accessed June 2020).

SCEC (2016). Seven steps to a disaster-resilient workplace. www.earthquakecountry.org/wp-content/themes/earthquakecountry_4_1/downloads/7_Steps_to_a_Disaster_Resilient_Workplace_FULL.pdf (accessed October 2020).

Shah, H. (2012). Catastrophe micro-insurance for those at the bottom of the pyramid. In: *Earthquake Engineering in Europe* (ed. M. Garevski), 549–561. Springer.

Smith, N. (2017). Better managing New Zealand's earthquake risks. https://www.beehive.govt.nz/speech/better-managing-new-zealand%E2%80%99s-earthquake-risks (accessed 15 September 2020).

Smyth, A.W., Altay, G., Deodatis, M. et al. (2004). Probabilistic benefit-cost analysis for eartquake damage mitigation: evaluation measures for apartment houses in Turkey. *Earthquake Spectra* 20: 171–204.

So, E. (2016). *Estimating Fatality Rates for Earthquake Loss Models*. Springer.

Solnit, R. (2009). *A Paradise Built in Hell*. New York: Penguin Books.

Spence, R., Oliveira, C., D'Ayala, D., Papa, F., and Zuccaro, G. (2000). The performance of strengthened masonry buildings in recent European earthquakes (paper 1366). *Presented at the 12th World Conference on Earthquake Engineering* (30 January–4 February 2000), Auckland.

Woo, G. (1999). *The Mathematics of Natural Catastrophes*. London: Imperial College Press.

Yanev, P. and Thompson, A. (2008). *Peace of Mind in Earthquake Country: How to Save Your Home, Business and Life*, 3e. San Francisco: Chronicle Books.

Index

Why Do Buildings Collapse in Earthquakes?: Building for Safety in Seismic Areas, First Edition. Robin Spence and Emily So.
© 2021 John Wiley & Sons Ltd. Published 2021 by John Wiley & Sons Ltd.